T0201425

METHOD OF LINES PDE ANALYSIS IN BIOMEDICAL SCIENCE AND ENGINEERING

METHOD OF LINES PDE ANALYSIS IN BIOMEDICAL SCIENCE AND ENGINEERING

WILLIAM E. SCHIESSER
Department of Chemical and Biomolecular Engineering
Lehigh University
Bethlehem, PA USA

Library of Congress Cataloging-in-Publication Data:

Names: Schiesser, W. E., author.
Title: Method of lines PDE analysis in biomedical science and engineering /
    William E. Schiesser.
Description: Hoboken, New Jersey : John Wiley & Sons, 2016. | Includes
    bibliographical references and index.
Identifiers: LCCN 2015043460 | ISBN 9781119130482
Subjects: LCSH: Differential equations, Partial–Numerical solutions. |
    Numerical analysis–Computer programs.
Classification: LCC QA377 .S35 2016 | DDC 610.1/515353–dc23 LC record available at
    http://lccn.loc.gov/2015043460

Typeset in 10/12pt, Times-Roman by SPi Global, Chennai, India.

Printed in the United States of America

10 9 8 7 6 5 4 3 2 1

To
Edward Amstutz
Glenn Christensen
Joseph Elgin
Harvey Neville

CONTENTS

PREFACE

The reporting of differential equation models in biomedical science and engineering (BMSE) continues at a remarkable pace. In this book, recently reported models based on initial-boundary-value ordinary and partial differential equations (ODE/PDEs) are described in chapters that have the following general format:

1. The model is stated as an ODE/PDE system, including the required initial conditions (ICs) and boundary conditions (BCs). The origin of PDEs based on mass conservation is discussed in Appendix A.

2. The coding (programming) of the model equations is presented as a series of routines in R, which primarily implements the method of lines (MOL) for PDEs. Briefly, in the MOL,

   - The partial derivatives in the spatial (boundary value) independent variables are replaced by approximations such as finite differences, finite elements, finite volumes, or spectral representations. In the present discussion, finite differences (FDs) are used, although alternatives are easily included.[1]

   - Derivatives with respect to an initial-value variable remain, which are expressed through a system of ODEs. An ODE solver (integrator) is then used to compute a numerical solution to the ODE/PDE system.[2]

3. The resulting numerical and graphical (plotted) solution is discussed and interpreted with respect to the model equations.

---

[1]Representative routines for the approximation of PDE spatial derivatives are listed in Appendix B.

[2]Additional ODEs, which might, for example, be BCs for PDEs, can naturally be included in the model MOL solution. The solution of a mixed ODE/PDE system is demonstrated in some of the BMSE applications. Also, differential algebraic equations (DAEs) can easily be included in a MOL solution through the use of a modified ODE solver or a DAE solver.

4. The chapter concludes with a review of the numerical (MOL) algorithm performance, general observations and results from the model, and possible extensions of the model.

The source of each model is included as one or more references. Generally, these are recent papers from the scientific and mathematics literature. Typically, a paper consists of some background discussion of the model, a statement of the model ODE/PDE system, presentation of a numerical solution of the model equations, and conclusions concerning the model and features of the solution.

What is missing in this format are the details of the numerical algorithms used to compute the reported solutions and the coding of the model equations. Also, the statement of the model is frequently incomplete such as missing equations, parameters, ICs, and BCs, so that reproduction of the solution with reasonable effort is virtually impossible.

We have attempted to address this situation by providing the source code of the routines, with a detailed explanation of the code, a few lines at a time. This approach includes a complete statement of the model (the computer will ensure this) and an explanation of how the numerical solutions are computed in enough detail that the reader can understand the numerical methods and coding and reproduce the solutions by executing the R routines (provided in a download).

In this way, we think that the formulation and use of the ODE/PDE models will be clear, including all of the mathematical details, so that readers can execute, then possibly experiment and extend, the models with reasonable effort. Finally, the intent of the detailed discussion is to explain the MOL formulation and methodology so that the reader can develop new ODE/PDE models and applications without becoming deeply involved in mathematics and computer programming.

In summary, the presentation is not as formal mathematics, for example, theorems and proofs. Rather, the presentation is by examples of recently reported ODE/PDE BMSE models including the details for computing numerical solutions, particularly documented source code. The author would welcome comments and suggestions for improvements (wes1@lehigh.edu).

WILLIAM E. SCHIESSER
Bethlehem, PA, USA
May 1, 2016

ABOUT THE COMPANION WEBSITE

This book is accompanied by a companion website:
www.wiley.com/go/Schiesser/PDE_Analysis

The website includes:

- Related R Routines

1

AN INTRODUCTION TO MOL ANALYSIS OF PDEs: WAVE FRONT RESOLUTION IN CHROMATOGRAPHY

This first chapter introduces a partial differential equation (PDE) model for chromatography which is a basic analytical method in biomedical science and engineering (BSME). For example, chromatography can be used to analyze a stream of various proteins through selective adsorption. Thus, the model can also be applied to adsorption as a basic procedure for separating biochemical species such as proteins. The computer implementation (programming, coding) of the model is in R[1].

The intent of this chapter is to

- Derive a basic chromatography PDE model, including the required initial conditions (ICs) and boundary conditions (BCs).
- Illustrate the coding of the model within the method of lines (MOL) through a series of R routines, including the use of library routines for integration of the PDE derivatives in time and space.
- Present the computed model solution in numerical and graphical (plotted) format.
- Discuss the features of the numerical solution and the performance of the algorithms used to compute the solution.
- Consider extensions of the model and the numerical algorithms.

[1]R is an open source scientific programming system that can be downloaded (at no cost) from the Internet. The R routines discussed in this book are available as a download from the author and the publisher.

Method of Lines PDE Analysis in Biomedical Science and Engineering, First Edition. William E. Schiesser.
© 2016 John Wiley & Sons, Inc. Published 2016 by John Wiley & Sons, Inc.
Companion website: www.wiley.com/go/Schiesser/PDE_Analysis

(1.1) 1D 2-PDE MODEL

The configuration of a chromatography column is illustrated in Fig. 1.1.

We can note the following details about the column represented in Fig. 1.1:

- The column is one dimensional (1D) with distance along the column, z, as the spatial (boundary value) independent variable. Time t is an initial value independent variable. A solid adsorbent is represented as spherical particles that fill the column. A fluid stream flows through the column in the interstices (voids) between the adsorbent particles. The flowing stream enters the base of the column at $z = 0$, and exits the top at $z = z_L$.
- The two PDE dependent variables are:
    - $u_1(z, t)$: concentration of the adsorbate (the chemical component to be processed) in the fluid stream.
    - $u_2(z, t)$: adsorbate concentration on the adsorbent.

    $u_1(z, t)$ and $u_2(z, t)$ are the PDE dependent variables. The PDEs that define these dependent variables are derived subsequently[2].
- The adsorbate enters the column at $z = 0$ with a prescribed (entering) concentration $u_{1e}(t)$ that serves as a boundary condition (BC) for the u_1 PDE[3]. Note that the boundary value can be a function of t.

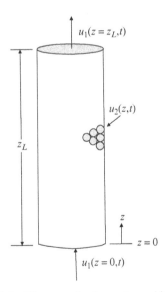

Figure 1.1   Diagram of a chromatographic column

---

[2]In accordance with the usual convention in PDE modeling, the dependent variables are designated with the letter u and a numerical subscript for each variable, e.g., $u_1(z, t)$, $u_2(z, t)$. The solution to the models PDEs will be the dependent variables in numerical form as a function of the independent variables, e.g., z, t.

[3]The term *boundary condition* follows from the use of a mathematical condition specified at the physical boundary of the system, in this case the adsorbate at $z = 0$, $u_1(z = 0, t) = u_{1e}(t)$.

- The exiting stream at $z = z_L$ has the concentration $u_1(z = z_L, t)$ which is a function of t. The t variation of this exiting stream is of primary interest when using the model. A plot of $u_1(z = z_L, t)$ against t is termed a *breakthrough curve*.
- An overall objective in formulating the model and computing numerical solutions is to determine $u_1(z = z_L, t)$, and in particular, how effective the chromatographic column is in alterting the entering stream with concentration $u_1(z = 0, t) = u_{1e}(t)$.

In summary, the numerical solution of the PDE model will give the dependent variables $u_1(z, t)$ and $u_2(z, t)$ as a function of z, t. $u_1(z = z_L, t)$ is a primary output from the model, that is, the outflow adsorbate concentration as a function of time.

A mass balance on the adsorbate stream[4] gives

$$\epsilon A \Delta z \frac{\partial u_1}{\partial t} = \epsilon A v u_1|_z - \epsilon A v u_1|_{z+\Delta z} - (1 - \epsilon) A \Delta z (k_f u_1 (u_2^e - u_2) - k_r u_2) \quad (1.1a)$$

where

u_1	concentration of adsorbate in the flowing stream
u_2	concentration of adsorbate on the adsorbent
u_2^e	equilibrium (saturation) concentration of adsorbate on the adsorbent
t	time
z	axial position along the column
A	cross sectional area of column (transverse to z)
v	superficial velocity of flow
ϵ	void fraction of the adsorbent interstices
k_f	forward rate constant for the adsorbate transfer from the fluid to the adsorbent
k_r	reverse rate constant for the adsorbate transfer from the adsorbent to the flowing stream

Table 1.1: Variables and parameters of eq. (1.1a)

Eq. (1.1a) is a mass conservation balance for the flowing adsorbate with the terms explained further in the following comments.

LHS-1: $\epsilon A \Delta z \dfrac{\partial u_1}{\partial t}$ - accumulation of adsorbate in the incremental volume $\epsilon A \Delta z$. The CGS units of this term are $(cm^2)(cm)(gmol/cm^3)(1/s) = gmol/s$, that is, the accumulation of adsorbate per second within the incremental volume $\epsilon A \Delta z$. If the derivative

---

[4] A 3D PDE is derived in Appendix A that can be reduced to eq. (1.1a) as a special case.

$\dfrac{\partial u_1}{\partial t}$ is negative, the adsorbate is depleted (reduced). Also, some elaboration of the units of length is possible.

- ϵ: $cm^3_{fluid}/cm^3_{column}$ (so that the void fraction is not dimensionless)
- A: cm^2_{column}
- Δz: cm_{column}
- u_1: g-mol/cm^3_{fluid}

Thus, more detailed units of the LHS t derivative of eq. (1.1a) are:

$$(cm^3_{fluid}/cm^3_{column})(cm^2_{column})(cm_{column})(g\text{-mol}/cm^3_{fluid})(1/s)=\text{gmol/s}$$

The distinction between cm_{fluid} and cm_{column} (and later, $cm_{adsorbent}$) will not generally be retained in the subsequent discussion (only cm will be used), but this distinction should be kept in mind when analyzing units in the model.

RHS-1: $\epsilon Avu_1|_z$ - flow (by convection) of absorbate into the incremental volume at z. The units of this term are $(cm^2)(cm/s)(gmol/cm^3)$ =gmol/s, that is, the flow of adsorbate per second into the incremental volume. Note that v has the units cm_{column}/s This is generally termed a *superficial* or *linear* velocity and is assumed constant across the chromatographic column (any wall effects are neglected).

RHS-2: $-\epsilon Avu_1|_{z+\Delta z}$ - flow (by convection) of absorbate out of the incremental volume at $z + \Delta z$. Again, the units of this term are $(cm^2)(cm/s)(gmol/cm^3)=$gmol/s, that is, the flow of adsorbate per second out of the incremental volume.

RHS-3: $-(1 - \epsilon)A\Delta z(k_f u_1(u_2^e - u_2) - k_r u_2)$ - volumetric rate of adsorption (when this term is negative, adsorbate moves from the fluid to the adsorbent) or desorption (when this term is positive). The units of this term are $(cm^2)(cm)(1/s)(gmol/cm^3)=$gmol/s, that is, the transfer of adsorbate per second within the incremental volume.

Three additional points about this term can be observed.

- u_2^e and u_2 are volumetric (not surface) adsorbent concentrations with the units gmol/ $cm^3_{absorbent}$. k_f has the units cm^3_{fluid}/gmol-s and k_r has the units 1/s (explained next).

  By definition,

$$cm^3_{fluid}+cm^3_{absorbent}=cm^3_{column}$$

and

$$(1\text{-}\epsilon)\rightarrow (1\text{-}cm^3_{fluid}/cm^3_{column}) = (cm^3_{column}\text{-}cm^3_{fluid})/cm^3_{column}$$
$$= cm^3_{adsorbent}/cm^3_{column}$$

Then the units of the term $-(1 - \epsilon)A\Delta z(k_f u_1(u_2^e - u_2) - k_r u_2)$ are:

$$(\text{cm}^3_{absorbent}/\text{cm}^3_{column})\,(\text{cm}^2_{column})(\text{cm}_{column})((\text{cm}^3_{fluid}/\text{gmol-s})$$
$$(\text{gmol}/\text{cm}^3_{fluid})(\text{gmol}/\text{cm}^3_{adsorbent})\text{-}(1/s)(\text{gmol}/\text{cm}^3_{adsorbent}))=\text{gmol/s}$$

- The forward rate of adsorption, $k_f u_1(u_2^e - u_2)$, is usually termed a *logistic* rate. Note that it is nonlinear from the product of the two dependent variables, $u_1 u_2$, which means that an analytical solution to the PDE model is probably precluded, but a numerical solution can be easily programmed and calculated. Also, for $(u_2^e - u_2) > 0$ this forward rate is positive giving adsorption from this term in eq. (1.1a), and for $(u_2^e - u_2) < 0$ this term reflects desorption (when the adsorbate concentration u_2 exceeds the equilibrium adsorbent concentration, u_2^e).
- When $(k_f u_1(u_2^e - u_2) - k_r u_2) > 0$, adsorption takes place (with a reduction in $\dfrac{\partial u_1}{\partial t}$ from eq. (1.1a) since this term is multiplied by a minus). Conversely, when $(k_f u_1(u_2^e - u_2) - k_r u_2) < 0$, this term reflects desorption (and an increase in $\dfrac{\partial u_1}{\partial t}$ from eq. (1.1a)).

If eq. (1.1a) is divided by $\epsilon A \Delta z$,

$$\frac{\partial u_1}{\partial t} = -\frac{vu_1|_{z+\Delta z} - vu_1|_z}{\Delta z} - \frac{(1-\epsilon)}{\epsilon}(k_f u_1(u_2^e - u_2) - k_r u_2)$$

or for $\Delta z \to 0$,

$$\frac{\partial u_1}{\partial t} = -\frac{\partial(vu_1)}{\partial z} - \frac{(1-\epsilon)}{\epsilon}(k_f u_1(u_2^e - u_2) - k_r u_2) \tag{1.1b}$$

Eq. (1.1b) is the PDE for the calculation of $u_1(z,t)$. For the subsequent analysis and programming, we will take v as independent of z so it can be taken outside the derivative in z (even though the transfer of adsorbate could affect v, but this will not be considered). v as a function of t is an interesting case that could be investigated through the use of eq. (1.1b). Note also that the column cross sectional area, A, canceled in going from eq. (1.1a) to eq. (1.1b), that is, we come to the somewhat unexpected conclusion that A does not appear in eq. (1.1b).

Also, in eq. (1.1b),

$$\frac{(1-\epsilon)}{\epsilon} \to \frac{\text{cm}^3_{adsorbent}/\text{cm}^3_{column}}{\text{cm}^3_{fluid}/\text{cm}^3_{column}} = \frac{\text{cm}^3_{adsorbent}}{\text{cm}^3_{fluid}}$$

as expected for consistent units in eq. (1.1b), that is, the units in the various terms in eq. (1.1b) are $\text{gmol}/\text{cm}^3_{fluid}$-s since eq. (1.1b) is a mass balance on the fluid.

A PDE for u_2 follows from an analogous mass balance for the adsorbent. The starting point is

$$(1-\epsilon)A\Delta z\frac{\partial u_2}{\partial t} = (1-\epsilon)A\Delta z(k_f u_1(u_2^e - u_2) - k_r u_2) \tag{1.2a}$$

Division by $(1 - \epsilon)A\Delta z$ gives

$$\frac{\partial u_2}{\partial t} = k_f u_1(u_2^e - u_2) - k_r u_2 \tag{1.2b}$$

Note that the adsorption terms in eqs. (1.1b) and (1.2b) are opposite in sign which indicates that the rate absorbate leaves (or enters) the fluid stream equals the rate adsorbate is transferred to (or leaves) the adsorbent. Also, the LHS and RHS terms in eq. (1.2b) have the units $\text{gmol/cm}^3_{adsorbent}$-s, since eq. (1.2b) is a mass balance on the adsorbent in an incremental volume $(1 - \epsilon)A\Delta z$. Again, A cancels in going from eq. (1.2a) to eq. (1.2b).

Eqs. (1.1b) and (1.2b) are a 2×2 (two equations in two unknowns) for the concentrations u_1, u_2. One other variable, u_2^e, appears in the adsorption rate in these two PDEs. This adsorbent equilibrium concentration might be assumed to be a constant, for example, corresponding to a monolayer of the adsorbate on the adsorbent. Or u_2^e can be considered a variable from an equilibrium relation such as, for example, a *Langmuir isotherm* of the form

$$u_2^e = \frac{c_1 u_1}{1 + c_2 u_1} \tag{1.3}$$

where c_1, c_2 are constants typically measured experimentally.

Eq. (1.1b) is first order in t and z (and is termed a *first-order, hyperbolic PDE*). Therefore, it requires one *initial condition* (IC)[5] and one *boundary condition* (BC).[6,7]

The IC is taken as

$$u_1(z, t = 0) = f_1(z) \tag{1.4a}$$

The BC is taken as

$$u_1(z = 0, t) = g_1(t) \tag{1.4b}$$

where $f_1(z)$ and $g_1(t)$ are prescibed functions of z and t, respectively.

Eq. (1.2b) is first order in t so it requires one IC

$$u_2(z, t = 0) = f_2(z) \tag{1.4c}$$

---

[5] An initial condition defines the value of the dependent variable, $u_1(z, t)$, for a particular value of the initial value independent variable, t, typically time in a physical application. t is defined over an open interval $t_0 \leq t \leq \infty$. The initial or beginning value, t_0, in the present case will be taken as $t_0 = 0$. Note that t can continue without limit.

[6] A boundary condition defines the value of the dependent variable, $u_1(z, t)$, for a particular value of the boundary value independent variable, z, typically at a physical boundary in an application. z can be defined over a finite interval, $z_0 \leq z \leq z_L$, a semi infinite interval, $z_0 \leq z \leq \infty$, or a fully infinite interval, $-\infty \leq z \leq \infty$. In the present case, we will use a finite interval corresponding to the length of the chromatographic column, $0 \leq z \leq z_L$ where z_L is the specified length of the column.

[7] In general, the number of required ICs equals the order of the highest order derivative in the initial value variable, one IC in the case of eq. (1.1b) for the first order derivative $\frac{\partial u_1}{\partial t}$. The number of required BCs equals the order of the highest order derivative in the boundary value variable, one BC in the case of eq. (1.1b) for the first order derivative $\frac{\partial u_1}{\partial z}$.

Eq. (1.4b) is a *Dirichlet* BC since the dependent variable u_1 is specified at the boundary $z = 0$. Other types of BCs are discussed in subsequent chapters.

Eqs. (1.1) to (1.4) constitute the PDE model for the chromatographic column. We next consider the programming of these equations within the MOL framework.

(1.2) MOL routines

The discussion of the routines for eqs. (1.1) to (1.4) starts with the main program.

(1.2.1) Main program

The listing of the main program follows.

```
#
# Delete previous workspaces
  rm(list=ls(all=TRUE))
#
#   1D, one component, chromatography model
#
#   The ODE/PDE system is
#
#   u1_t = -v*u1_z - (1 - eps)/eps*rate              (1.1b)
#
#   u2_t = rate                                      (1.2b)
#
#   rate = kf*u1*(u2eq - u2) - kr*u2
#
#   u2eq = c1*u1/(1 + c2*u1)                          (1.3)
#
#   Boundary condition
#
#     u1(z=0,t) = step(t)                             (1.4b)
#
#   Initial conditions
#
#     u1(z,t=0) = 0                                   (1.4a)
#
#     u2(z,t=0) = 0                                   (1.4c)
#
#   The method of lines (MOL) solution for eqs. (1.1) to
#   (1.4) is coded below.  Specifically, the spatial
#   derivative in the fluid balance, u1_z in eq. (1.1b),
#   is replaced by one of four approximations as selected
#   by the variable ifd.
#
# Access ODE integrator
```

```r
  library("deSolve");
#
# Access files
  setwd("g:/chap1");
  source("pde_1.R") ;source("step.R")   ;
  source("dss004.R");source("dss012.R");
  source("dss020.R");source("vanl.R")   ;
  source("max3.R")   ;
#
# Step through cases
  for(ncase in 1:2){
#
#   Model parameters
     v=1; eps=0.4; u10=0; u20=0;
    c1=1;    c2=1; zL=50;   n=41;
    if(ncase==1){ kf=0; kr=0; }
    if(ncase==2){ kf=1; kr=1; }
#
# Select an approximation for the convective derivative u1z
#
#   ifd = 1: Two point upwind approximation
#
#   ifd = 2: Centered approximation
#
#   ifd = 3: Five point, biased upwind approximation
#
#   ifd = 4: van Leer flux limiter
#
  ifd=1;
#
# Level of output
#
#   Detailed output   - ip = 1
#
#   Brief (IC) output - ip = 2
#
  ip=1;
#
# Initial condition
  u0=rep(0,2*n);
  for(i in 1:n){
      u0[i]=u10;
    u0[i+n]=u20;
  }
  t0=0;tf=150;nout=51;
  tout=seq(from=t0,to=tf,by=(tf-t0)/(nout-1));
```

```
  ncall=0;
#
# ODE integration
  out=ode(func=pde_1,times=tout,y=u0);
#
# Store solution
  u1=matrix(0,nrow=nout,ncol=n);
  u2=matrix(0,nrow=nout,ncol=n);
  t=rep(0,nout);
  for(it in 1:nout){
  for(iz in 1:n){
    u1[it,iz]=out[it,iz+1];
    u2[it,iz]=out[it,iz+1+n];
  }
    t[it]=out[it,1];
  }
#
# Display ifd, ncase
  cat(sprintf("\n ifd = %2d   ncase = %2d",ifd,ncase));
#
# Display numerical solution
  if(ip==1){
  cat(sprintf(
    "\n\n      t      u1(z=zL,t)  rate(z=zL,t)\n"));
  u2eq=rep(0,nout);rate=rep(0,nout);
  for(it in 1:nout){
    u2eq[it]=c1*u1[it,n]/(1+c2*u1[it,n]);
    rate[it]=kf*u1[it,n]*(u2eq[it]-u2[it,n])-kr*u2[it,n];
    cat(sprintf(
      "%7.2f%12.4f%12.4f\n",t[it],u1[it,n],rate[it]));
  }
  }
#
# Store solution for plotting
  u1plot=rep(0,nout);tplot=rep(0,nout);
  for(it in 1:nout){
    u1plot[it]=u1[it,n];
     tplot[it]=t[it];
  }
#
# Calls to ODE routine
  cat(sprintf("\n ncall = %4d\n",ncall));
#
# Plot for u1(z=zL,t)
# ncase = 1
  if(ncase==1){
```

```
  par(mfrow=c(1,1))
  plot(tplot,u1plot,xlab="t",ylab="u1(z=zL,t)",
    lwd=2,main="u1(z=zL,t) vs t, ncase=1\n
    line - anal, o - num",type="l",
    xlim=c(0,tplot[nout]));#,ylim=c(0,1));
  points(tplot,u1plot, pch="o",lwd=2);
  }
#
# Analytical solution, ncase=1
  if(ncase==1){
  u1expl=rep(0,nout);
  for(it in 1:nout){
    u1expl[it]=step(tplot[it],zL,v);
  }
  lines(tplot,u1expl,lwd=2,type="l");
  }
#
# ncase = 2
  if(ncase==2){
  par(mfrow=c(1,1))
  plot(tplot,u1plot,xlab="t",ylab="u1(z=zL,t)",
    lwd=2,main="u1(z=zL,t) vs t, ncase=2",
    type="l",xlim=c(0,tplot[nout]));#,ylim=c(0,1));
  points(tplot,u1plot, pch="o",lwd=2);
  }
#
# Next case
  }
```

Listing 1.1: Main program pde_1_main for eqs. (1.1) to (1.4)

We can note the following details about pde_1_main.

- Previous files are cleared and a series of documentation comments for the ODE/PDE system is included that restates eqs. (1.1) to (1.4) in the text.

```
#
# Delete previous workspaces
  rm(list=ls(all=TRUE))
#
#    1D, one component, chromatography model
#
#    The ODE/PDE system is
#
#    u1_t = -v*u1_z - (1 - eps)/eps*rate          (1.1b)
#
```

```
#    u2_t = rate                                              (1.2b)
#
#    rate = kf*u1*(u2eq - u2) - kr*u2
#
#    u2eq = c1*u1/(1 + c2*u1)                                 (1.3)
#
#    Boundary condition
#
#      u1(z=0,t) = step(t)                                    (1.4b)
#
#    Initial conditions
#
#      u1(z,t=0) = 0                                          (1.4a)
#
#      u2(z,t=0) = 0                                          (1.4c)
#
#    The method of lines (MOL) solution for eqs. (1.1) to
#    (1.4) is coded below.  Specifically, the spatial
#    derivative in the fluid balance, u1_z in eq. (1.1b),
#    is replaced by one of four approximations as selected
#    by the variable ifd.
```

The IC and BC functions of eqs. (1.4), $f_1(z) = 0, g_1(t) = h(t), f_2(z) = 0$, are explained subsequently ($h(t)$ is the *unit step function* or *Heaviside function*).

- The R ODE integrator library deSolve and a series of routine discussed subsequently are accessed.

```
#
# Access ODE integrator
  library("deSolve");
#
# Access files
  setwd("g:/chap1");
  source("pde_1.R") ;source("step.R")   ;
  source("dss004.R");source("dss012.R");
  source("dss020.R");source("van1.R")   ;
  source("max3.R")   ;
```

The set working directory, setwd, will have to be edited for the local computer. Note the forward slash, /, rather than the usual backslash, \. The source utility is used to select individual files that make up the complete code for the model of eqs. (1.1) to (1.4). These files are explained subsequently.

- A for is used to step through a series of (two) cases, ncase=1,2.

```
#
# Step through cases
```

```
  for(ncase in 1:2){
#
#   Model parameters
      v=1; eps=0.4; u10=0; u20=0;
      c1=1;    c2=1; zL=50;   n=41;
      if(ncase==1){ kf=0; kr=0; }
      if(ncase==2){ kf=1; kr=1; }
```

The parameters in eqs. (1.1) to (1.4) for each case are defined numerically. In particular, for ncase=1, no adsorption takes place so that the fluid with adsorbate concentration $u_1(z,t)$ merely flows through the column and there is no up take of adsorbate with concentration $u_2(z,t)$ onto the adsorbent. This special condition is used to check the coding of the model as discussed subsequently. For ncase=2, the effect of adsorbate transfer to the adsorbent can be observed in the fluid outlet with concentration $u_1(z = z_L, t)$.

- An approximation for the spatial derivative $\dfrac{\partial u_1}{\partial z}$ in eq. (1.1b) (v constant and therefore outside of the derivative) is selected with index ifd. The performance of the four approximations is discussed subsequently.

```
#
# Select an approximation for the convective derivative u1z
#
#    ifd = 1: Two point upwind approximation
#
#    ifd = 2: Centered approximation
#
#    ifd = 3: Five point, biased upwind approximation
#
#    ifd = 4: van Leer flux limiter
#
   ifd=1;
```

- A level of numerical output is selected with ip. Initially, ip=1 is used to give detailed numerical output along with graphical (plotted) output. ip=2 can be used to give only graphical output (when experimenting with the model).

```
#
# Level of output
#
#    Detailed output   - ip = 1
#
#    Brief (IC) output - ip = 2
#
   ip=1;
```

- ICs (1.4a) and (1.4c) are programmed as zero (*homogeneous*) ICs (since u10 = u20 = 0). Note that these ICs are placed in a single vector or 1D array u0 as required by the ODE integrator ode discussed next. This vector is first declared (allocated, sized) with the utility rep (2*n = 2*41 = 82 zero elements).

```
#
# Initial condition
  u0=rep(0,2*n);
  for(i in 1:n){
      u0[i]=u10;
    u0[i+n]=u20;
  }
  t0=0;tf=150;nout=51;
  tout=seq(from=t0,to=tf,by=(tf-t0)/(nout-1));
  ncall=0;
```

The time scale is defined as $0 \leq t \leq 150$ with nout=51 points in t for the numerical solution. The utility seq is used to define the 51 values $t = 0, 3, 6, ..., 150$. Finally, the counter for the calls to the ODE routine pde_1 is initialized. The use of this counter is discussed later.

- The 2*n = 82 ODEs are integrated numerically with the library integrator ode (part of the deSolve library specified previously).

```
#
# ODE integration
  out=ode(func=pde_1,times=tout,y=u0);
```

The input arguments for ode require some explanation.

- The routine for the MOL/ODEs that approximate PDEs (1.1b), (1.2b), pde_1, is declared for the parameter func (which is a reserved argument name). func does not have to be the first input argument, but by convention, it usually is when calling one of the R integrators (ode in this case).
- The vector of output values of t, tout, (defined previously) is assigned to the input argument times. Again, times is a reserved name and can be placed anywhere in the input argument list.
- The IC vector, u0, is assigned to the parameter y. The length of this IC vector tells ode how many ODEs are to be integrated, in this case 2*n = 82. Note that the number of ODEs is not specified explicitly in the input argument list.

Numerical solutions to eqs. (1.1) to (1.4) are returned by ode in the 2D array out. The content of this solution array is explained next. The various ODE integrators in deSolve generally follow this format.

- The numerical solution is placed in matrices and a vector. These arrays are first declared with the utilities matrix (for u_1, u_2 of eqs. (1.1b), (1.2b)) and rep (for t of eqs. (1.1b) and (1.2b)).

```
#
# Store solution
  u1=matrix(0,nrow=nout,ncol=n);
  u2=matrix(0,nrow=nout,ncol=n);
  t=rep(0,nout);
  for(it in 1:nout){
  for(iz in 1:n){
    u1[it,iz]=out[it,iz+1];
    u2[it,iz]=out[it,iz+1+n];
  }
    t[it]=out[it,1];
  }
```

A pair of nested fors is used to place the numerical solutions in u1,u2. The outer for with index it steps through t for $0 \le t \le 150$. The inner for with index iz steps through z for $0 \le z \le z_L$ with $z_L = 50$ (defined previously) and a spatial increment $(50 - 0)/(41 - 1) = 1.25$, that is, $z = 0, 1.25, 2.50, ..., 50$ (based on n=41 points in z).

The solution array out has the dimensions out(nout,2*n+1) = out(51,82+1), that is, 82 ODEs at nout=51 points in t (including $t = 0$). The offset of 1 in iz+1,iz+1+n,2*n+1,82+1 reflects the additional space for t, so that out[it,1] contains the 51 values of t. This ordering of the output array out is a unique feature of the ODE integrators in deSolve, including ode.

- The index for the spatial differentiator, ifd, and the value of ncase are displayed.

```
#
# Display ifd, ncase
  cat(sprintf("\n ifd = %2d   ncase = %2d",ifd,ncase));
```

- For ip=1, the numerical solution is displayed as (1) t, (2) $u_1(z = z_L, t)$, and (3) $rate(t) = k_f u_1(z = z_L, t)(u_2^e - u_2(z = z_L, t)) - k_r u_2(z = z_L, t)$ (note the use of n for $z = z_L$). Vectors are first defined with the utility rep.

```
#
# Display numerical solution
  if(ip==1){
  cat(sprintf(
    "\n\n     t       u1(z=zL,t)   rate(z=zL,t)\n"));
  u2eq=rep(0,nout);rate=rep(0,nout);
  for(it in 1:nout){
    u2eq[it]=c1*u1[it,n]/(1+c2*u1[it,n]);
    rate[it]=kf*u1[it,n]*(u2eq[it]-u2[it,n])-kr*u2[it,n];
    cat(sprintf(
      "%7.2f%12.4f%12.4f\n",t[it],u1[it,n],rate[it]));
  }
  }
```

u_2^e = u2eq is computed from the isotherm of eq. (1.3).

- $u_1(z = z_L, t)$, t are stored for subsequent plotting.

```
#
# Store solution for plotting
  u1plot=rep(0,nout);tplot=rep(0,nout);
  for(it in 1:nout){
    u1plot[it]=u1[it,n];
     tplot[it]=t[it];
  }
```

- At the end of the solution (after the call to ode), the number of calls to the MOL/ODE routine pde_1 is displayed (this routine is discussed next).

```
#
# Calls to ODE routine
  cat(sprintf("\n ncall = %4d\n",ncall));
```

- $u_1(z = z_L, t)$ is plotted against t for ncase=1 (no adsorption). For this case, an analytical solution is available that is plotted as a solid line while the numerical solution is plotted as points on a solid line (this is clear in Fig. 1.2).

```
#
# Plot for u1(z=zL,t)
# ncase = 1
  if(ncase==1){
  par(mfrow=c(1,1))
  plot(tplot,u1plot,xlab="t",ylab="u1(z=zL,t)",
    lwd=2,main="u1(z=zL,t) vs t, ncase=1\n
    line - anal, o - num",type="l",
    xlim=c(0,tplot[nout]));#,ylim=c(0,1));
  points(tplot,u1plot, pch="o",lwd=2);
  }
```

The scaling of the y axis is deactivated as a comment, #,ylim=c(0,1)); so that oscillations in the solution outside $0 \leq u_1(z = z_L, t) \leq 1$ can be accommodated with the default scaling for the y axis (the oscillations are a numerical artifact that is an incorrect part of the numerical solution as discussed subsequently).

- For ncase=1 (no adsorption), the analytical solution to eq. (1.1b) is computed by a call to step (as explained subsequently). The resulting plot of the analytical solution is superimposed on the preceding plot of $u_1(z = z_L, t)$ (see Fig. 1.2).

```
#
# Analytical solution, ncase=1
  if(ncase==1){
  u1expl=rep(0,nout);
  for(it in 1:nout){
```

```
    u1expl[it]=step(tplot[it],zL,v);
    }
    lines(tplot,u1expl,lwd=2,type="l");
    }
```

- For ncase=2 (with adsorption), the numerical solution $u_1(z = z_L, t)$ is plotted against t as points on a solid line.

```
#
# ncase = 2
  if(ncase==2){
  par(mfrow=c(1,1))
  plot(tplot,u1plot,xlab="t",ylab="u1(z=zL,t)",
    lwd=2,main="u1(z=zL,t) vs t, ncase=2",
    type="l",xlim=c(0,tplot[nout]));#,ylim=c(0,1));
  points(tplot,u1plot, pch="o",lwd=2);
  }
#
# Next case
  }
```

The final } concludes the for in ncase.
The ODE routine pde_1 called by ode (Listing 1.1) is considered next.

(1.2.2) MOL/ODE routine

The ODE routine pde_1 called by ode (Listing 1.1) is in Listing 1.2.

```
  pde_1=function(t,u,parms){
#
# Function pde_1 computes the t derivative vector of the u vector
#
# One vector to two PDEs
  u1=rep(0,n);u2=rep(0,n);
  for (i in 1:n){
    u1[i]=u[i];
    u2[i]=u[i+n];
  }
#
# Boundary condition
  u1[1]=step(t,0,v);
#
# First order spatial derivative
#
#    ifd = 1: Two point upwind finite difference (2pu)
     if(ifd==1){ u1z=dss012(0,zL,n,u1,v); }
#
```

```
#     ifd = 2: Three point center finite difference (3pc)
      if(ifd==2){ u1z=dss004(0,zL,n,u1); }
#
#     ifd = 3: Five point biased upwind approximation (5pbu)
      if(ifd==3){ u1z=dss020(0,zL,n,u1,v); }
#
#     ifd = 4: van Leer flux limiter
      if(ifd==4){ u1z=vanl(0,zL,n,u1,v); }
#
# Temporal derivatives, mass transfer rate
    u1t=rep(0,n); u2t=rep(0,n);
  u2eq=rep(0,n);rate=rep(0,n);
#
#     u1t, u2t
      for(i in 1:n){
        u2eq[i]=c1*u1[i]/(1+c2*u1[i]);
        rate[i]=kf*u1[i]*(u2eq[i]-u2[i])-kr*u2[i];
        if(i==1){
          u1t[i]=0;
        }else{
          u1t[i]=-v*u1z[i]-(1-eps)/eps*rate[i];
        }
          u2t[i]=rate[i];
      }
#
# Two PDEs to one vector
  ut=rep(0,2*n);
  for(i in 1:n){
      ut[i]=u1t[i];
    ut[i+n]=u2t[i];
  }
#
# Increment calls to pde_1
  ncall<<-ncall+1;
#
# Return derivative vector
  return(list(c(ut)));
}
```

Listing 1.2: ODE routine pde_1 for eqs. (1.1) to (1.4)

We can note the following points about pde_1.

- The function is defined.

  ```
  pde_1=function(t,u,parms){
  #
  # Function pde_1 computes the t derivative vector of the u
      vector
  ```

The input argument t is the current value of t along the numerical solution. u is the current vector of (82) ODE dependent variables. parm is a set of input parameters for the ODEs; in this case it is unused (but is required in the input arguments). Note that the input arguments are not assigned to reserved names (as in the call to ode in Listing 1.1), so the order that they are specified must be maintained, e.g., t first followed by u.

- The single vector u that is the second input argument to pde_1 is placed in two vectors, u1 for eq. (1.1b) and u2 for eq. (1.2b). This is not a required step, but rather, is used so that the subsequent programming can be in terms of variables closely resembling u_1, u_2 in eqs. (1.1b), (1.2b).

```
#
# One vector to two PDEs
  u1=rep(0,n);u2=rep(0,n);
  for (i in 1:n){
    u1[i]=u[i];
    u2[i]=u[i+n];
  }
```

The two vectors u1,u2 are first declared with the rep utility.

- The BC for eq. (1.1b), eq. (1.4b) with $u_1(z = 0, t) = h(t) = $ u1(1), is specified as a unit step in function step (discussed subsequently).

```
#
# Boundary condition
  u1[1]=step(t,0,v);
```

The arguments of step are for the current value of t, the value $z = 0$, and the velocity v in eq. (1.1b) (and numerically defined previously).

- The first derivatives in z in eq. (1.1b), $\dfrac{\partial u_1}{\partial z}$, is computed by one of four spatial differentiators as selected by ifd (set previously). Some of the details of these differentiators and their performance as evaluated by comparison of the numerical solution with the analytical solution (for ncase=1) are considered subsequently.

```
#
# First order spatial derivative
#
#    ifd = 1: Two point upwind finite difference (2pu)
     if(ifd==1){ u1z=dss012(0,zL,n,u1,v); }
#
#    ifd = 2: Three point center finite difference (3pc)
     if(ifd==2){ u1z=dss004(0,zL,n,u1); }
#
#    ifd = 3: Five point biased upwind approximation (5pbu)
     if(ifd==3){ u1z=dss020(0,zL,n,u1,v); }
```

```
#
#    ifd = 4: van Leer flux limiter
     if(ifd==4){ u1z=vanl(0,zL,n,u1,v); }
```

- Vectors for the LHS derivatives in t in eqs. (1.1b), (1.2b), the equilibrium concentration u_1^e, and the rate of adsorption in eqs. (1.1b), (1.2b), are declared with the rep utility over the n=41 points in z.

```
#
# Temporal derivatives, mass transfer rate
    u1t=rep(0,n); u2t=rep(0,n);
    u2eq=rep(0,n);rate=rep(0,n);
```

- The LHS derivatives in t in eqs. (1.1b), (1.2b), $\dfrac{\partial u_1}{\partial t}, \dfrac{\partial u_2}{\partial t}$, are programmed in a for over the n=41 points in z.

```
#
#    u1t, u2t
     for(i in 1:n){
        u2eq[i]=c1*u1[i]/(1+c2*u1[i]);
        rate[i]=kf*u1[i]*(u2eq[i]-u2[i])-kr*u2[i];
        if(i==1){
          u1t[i]=0;
        }else{
          u1t[i]=-v*u1z[i]-(1-eps)/eps*rate[i];
        }
          u2t[i]=rate[i];
     }
```

We can note the following details in this programming.
- The equilibrium concentration u_2^e in eq. (1.3) is programmed first.
  ```
  u2eq[i]=c1*u1[i]/(1+c2*u1[i]);
  ```
- The adsorption rate in eqs. (1.1b) and (1.2b) is then programmed.
  ```
  rate[i]=kf*u1[i]*(u2eq[i]-u2[i])-kr*u2[i];
  ```
- The 41 MOL/ODEs are programmed.
  ```
          if(i==1){
            u1t[i]=0;
          }else{
            u1t[i]=-v*u1z[i]-(1-eps)/eps*rate[i];
          }
            u2t[i]=rate[i];
       }
  ```
Since $u_1(z = 0, t)$ is specified through BC (1.4b), its derivative in t is set to zero so that the ODE integrator, ode, will not move it away from its prescribed BC value, that is, u1t[i]=0;. Otherwise, eq. (1.1b) is programmed as
```
u1t[i]=-v*u1z[i]-(1-eps)/eps*rate[i];
```

Eq. (1.2b) is programmed as

```
u2t[i]=rate[i];
```

The close resemblance of this programming to the PDEs, eqs. (1.1b), (1.2b), is one of the principal advantages of the MOL.

- The two derivatives vectors, u1t,u2t, are placed in a single vector ut (of length $2*n = 2*41 = 82$) to be returned to the ODE integrator ode called in the main program of Listing 1.1.

```
#
# Two PDEs to one vector
  ut=rep(0,2*n);
  for(i in 1:n){
      ut[i]=u1t[i];
    ut[i+n]=u2t[i];
  }
```

The derivative vector ut is first declared with a rep.

- The counter for the calls to pde_1 is incremented and its value is returned to the main program of Listing 1.1 with <<-.

```
#
# Increment calls to pde_1
  ncall<<-ncall+1;
#
# Return derivative vector
  return(list(c(ut)));
}
```

The derivative vector ut is returned to ode as a list which is a requirement of ode (and generally, the ODE integrators in deSolve). c() is the vector operator in R.
      The final } concludes function pde_1.

Additional subordinate routines called in the preceding program are now considered.

(1.2.3) Subordinate routines

The unit step (Heaviside function) $h(t)$ in BC (1.4b) with $g_1(t) = h(t)$ is programmed in step.

```
step=function(t,z,v) {
#
# Function step approximates a unit step function
#
  tzv=t-z/v;
```

```
  if(tzv <0){u1s=0;  }
  if(tzv >0){u1s=1;  }
  if(tzv==0){u1s=0.5;}
#
# Return step
  return(c(u1s));
}
```

Listing 1.3: Function step for a unit step

We can note the following details about Listing 1.3.

- The unit step is a traveling wave with argument tzv=t-z/v[8]. The step is a finite discontinuity that occurs at $t - z/v = 0$. This discontinuity, which occurs at $z = 0$ in BC (1.4b), is approximated by three ifs. For $tzv < 0$, the function is 0, for $tzv > 0$, the function is 1, and for $tzv = 0$, the function is 0.5. This approximation is required since the unit step at $t - z/v = 0$ is undefined (it is not single valued). Analysis of the unit step and the associated solution of eqs. (1.1) to (1.4) is given subsequently.

- The value of the function, u1s, is returned as a 1-vector (through the operator c()).

The form of the unit step from function step will be clear from the graphical output of the numerical solutions of eqs. (1.1) to (1.4), e.g., in Fig. 1.2. In summary, step gives an approximation to a unit step as output for a given point along the chromatographic column z as a function of time t. This traveling unit step will be clear from the subsequent solution of eqs. (1.1) to (1.4).

The other routines called in the preceding programming, that is, dss004 to max3 as accessed by a source in the main program of Listing 1.1, are library routines and are briefly discussed later when the numerical solutions to eqs. (1.1) to (1.4) are considered. The source code for these routines is available from a download site (see the publisher's Web site for this book).

We next consider the output from the R routines in Listings 1.1, 1.2 and 1.3.

(1.3) Model output, single component chromatography

We first consider FDs as implemented with ifd = 1,2,3

(1.3.1) FDs, step BC

Abbreviated numerical output from the execution of the routines in Listings 1.1, 1.2 and 1.3 is given in Table 1.2.

---

[8] A detailed discussion of traveling wave solutions to PDEs is given in [1].

```
ifd =  1    ncase =  1

   t       u1(z=zL,t)   rate(z=zL,t)
 0.00        0.0000        0.0000
 3.00        0.0000        0.0000
               .             .
               .             .
               .             .
 (output for t = 6 to 21 deleted)
               .             .
               .             .
               .             .
24.00        0.0000        0.0000
27.00        0.0003        0.0000
30.00        0.0017        0.0000
33.00        0.0081        0.0000
36.00        0.0277        0.0000
39.00        0.0728        0.0000
42.00        0.1543        0.0000
45.00        0.2737        0.0000
48.00        0.4194        0.0000
51.00        0.5708        0.0000
54.00        0.7073        0.0000
57.00        0.8158        0.0000
60.00        0.8927        0.0000
63.00        0.9420        0.0000
66.00        0.9708        0.0000
69.00        0.9862        0.0000
72.00        0.9939        0.0000
75.00        0.9975        0.0000
78.00        0.9990        0.0000
81.00        0.9996        0.0000
84.00        0.9999        0.0000
87.00        1.0000        0.0000
90.00        1.0000        0.0000
               .             .
               .             .
               .             .
      (output for t = 93 to
          144 deleted)
               .             .
               .             .
               .             .
147.00       1.0000        0.0000
150.00       1.0000        0.0000
```

```
ncall =   609

ifd =  1    ncase =  2

    t       u1(z=zL,t)   rate(z=zL,t)
  0.00        0.0000        0.0000
  3.00        0.0000        0.0000
                 .             .
                 .             .
                 .             .
  (output for t = 6 to 24 deleted)
                 .             .
                 .             .
                 .             .
 27.00        0.0001        0.0000
 30.00        0.0008        0.0000
 33.00        0.0031        0.0000
 36.00        0.0089        0.0000
 39.00        0.0196        0.0001
 42.00        0.0359        0.0003
 45.00        0.0577        0.0007
 48.00        0.0852        0.0012
 51.00        0.1185        0.0018
 54.00        0.1584        0.0027
 57.00        0.2057        0.0038
 60.00        0.2617        0.0050
 63.00        0.3269        0.0063
 66.00        0.4014        0.0075
 69.00        0.4839        0.0084
 72.00        0.5710        0.0087
 75.00        0.6580        0.0084
 78.00        0.7393        0.0075
 81.00        0.8101        0.0062
 84.00        0.8676        0.0048
 87.00        0.9114        0.0035
 90.00        0.9430        0.0024
 93.00        0.9646        0.0016
 96.00        0.9787        0.0010
 99.00        0.9875        0.0006
102.00        0.9929        0.0004
105.00        0.9961        0.0002
108.00        0.9979        0.0001
111.00        0.9989        0.0001
114.00        0.9994        0.0000
117.00        0.9997        0.0000
120.00        0.9999        0.0000
```

123.00	0.9999	0.0000
126.00	1.0000	0.0000
129.00	1.0000	0.0000
132.00	1.0000	0.0000
135.00	1.0000	0.0000
138.00	1.0000	0.0000
141.00	1.0000	0.0000
144.00	1.0000	0.0000
147.00	1.0000	0.0000
150.00	1.0000	0.0000

```
ncall = 1143
```

Table 1.2: Selected numerical output for eqs. (1.1) to (1.4) from pde_1_main and pde_1 for ncase=1,2

We can note the following details about Table 1.2.

- The output is for $0 \leq t \leq 150$ as programmed in Listing 1.1.
- The ICs of eqs. (1.4a) and (1.4c) are confirmed (checking the ICs of a numerical solution is always a good idea since if the ICs are incorrect, the solution will start incorrectly and will therefore most likely be entirely incorrect).
- The rate of adsorption for ncase=1 is zero as expected since no adsorption takes place ($rate = k_f = k_r = 0$ in eqs. (1.1b), (1.2b)).
- The rate of adsorption for ncase=2 goes through a maximum of 0.0087 at $t = 72$. This maximum is expected as the adsorption on the adsorbate increases initially, then reaches a point where desorption begins to reduce the rate. Eventually, the rate returns to zero. Also, the maximum adsorption rate occurs at approximately the half way point in the transient, $u_1(z = z_L, t = 72) = 0.5710$, where the rate of change of u_1 is greatest (see Fig. 1.3).
- The solution is smooth (e.g., no oscillations) and approaches the expected final value, $u_1(z = z_L, t) = 1$. In other words, the two-point upwind finite difference (FD) approximation in dss012 corresponding to ifd=1 appears to function quite well. However, there is a significant error as reflected in Fig. 1.2 where the exact and analytical solutions are compared. This error is discussed in more detail later.
- The accuracy of a numerical PDE solution generally cannot be determined directly or explicitly (since this implies an analytical solution is available for computing the exact error). For ncase=1, an analytical solution is available, but this is unusual. In fact, numerical solutions are generally computed because analytical solutions are not available.

  However, each new numerical solution should be evaluated indirectly. Here are three procedures for an approximate error analysis that do not require an analytical solution for the full PDE problem.

  - The accuracy of the integration in z could be inferred by: (1) observing the effect of increasing the number of grid points, e.g., n=41 to n=81 (termed h

refinement since the grid spacing, which is typically given the symbol h, is varied when the number of grid points is changed), and (2) by changing the spatial derivative approximation, e.g., another value of ifd (termed *p refinement* since the order of the FD approximation[9], which is typically given the symbol p, is varied as the FD approximation is changed). The procedure then in (1) and (2) is to change h and p and observe the effect on the numerical solution. For example, if the solution does not change in the third figure, three-figure accuracy can be inferred. However, this is not a proof of numerical accuracy. Rather, it is just an estimate of accuracy.

– The accuracy of the integration in t could be inferred by changing the error tolerances specified for the ODE integrator. In the case of the R integrator ode (called in Listing 1.1), default error tolerances of 1×10^{-6} (absolute and relative) are used in ode unless the defaults are reset to other values before ode is called. ode is a sophisticated ODE integrator that changes the integration interval (h refinement) and the algorithm order (p refinement) in an attempt to meet the default or user-specified error tolerances. If ode is unable to meet the error tolerances, it issues a warning message to this effect[10].

– Special case analytical solutions can be used to test the numerical solutions. This is particularly useful in locating coding (programming) errors. For example, for ncase=1, the available analytical solution can be compared with the numerical solution as in Fig. 1.2.

• The total calls to pde_1 is ncall = 609 for ncase=1 and ncall = 1143 for ncase=2 indicating that ode computed numerical solutions with modest computational effort. The solutions presumably have the accuracy indicated through h and p refinement (as discussed subsequently).

Additional features of the numerical solution are evident in Figs. 1.2 and 1.3. We can note the following details in Fig. 1.2:

• The analytical solution, plotted as a solid line, is an approximation to a unit step at $t = 50$. This solution is derived in the following way.

---

[9]The order of a FD approximation refers to the power p in a formula of the form

$$error = c_1 \Delta z^p$$

where *error* is the *truncation error*, c_1 is a constant, Δz is the grid spacing in z and p is the order of the FD approximation. This formula for the error suggests that the error decreases with decreasing Δz (as the number of grid points in z is increased). For the two-point upwind FD approximations in dss012, $p = 1$ so the truncation error varies linearly with Δz and they are termed *first-order correct*. The term truncation error refers to the error resulting from truncating the Taylor series from which the FD is derived. For the five-point FD approximations in dss004 (with idf=2), $p = 4$ and for the five-point FD approximations in dss020 (with ifd=3), $p = 4$. These FD approximations of various orders are discussed subsequently.

[10]The adjustment of the FD integration intervals in ode is also termed *r refinement* where the r designates automatic (algorithmic, adaptive) refinement of the grid. This term is generally applied to the refinement of spatial grids. However, the spatial grids in dss004, dss012, dss020 are not refined automatically, i.e., they are fixed or constant grids that can be h-refined by changing the number of spatial grid points.

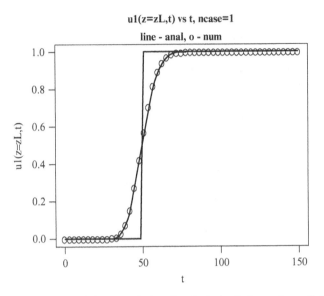

Figure 1.2   Comparison of the numerical and analytical solutions of eqs. (1.1b), (1.2b)

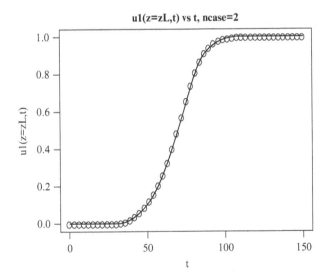

Figure 1.3   Numerical solution of eqs. (1.1b), (1.2b), `ncase=2`, `ifd=1`

Eq. (1.1b) for $k_f = k_r = 0$, v constant (`ncase=1` in Listing 1.1) is the *linear advection equation*,

$$\frac{\partial u_1}{\partial t} = -v\frac{\partial u_1}{\partial z} \tag{1.5a}$$

with IC (1.4a) and BC (1.4b). For the present analysis, we take (in eq. (1.4a))

$$f_1(z) = 0 \qquad (1.5b)$$

and (in eq. (1.4b))

$$g_1(t) = h(t) \qquad (1.5c)$$

where $h(t)$ is the unit step (Heaviside) function

$$h(t) = \begin{cases} 0, & t < 0 \\ 1, & t > 0 \end{cases} \qquad (1.5d)$$

Note that $h(t)$ is not defined at $t = 0$.

If we assume a solution to eq. (1.5a) of the form

$$u_1(z,t) = g_1(t - z/v) = g_1(\lambda); \quad \lambda = t - z/v \qquad (1.6a)$$

with $g_1(\lambda < 0) = 0$ (recall g_1 is the BC function of eq. (1.4b)), substitution of eq. (1.6a) in eq. (1.5a) gives

$$\frac{\partial u_1}{\partial t} = \frac{dg_1}{d\lambda}\frac{\partial \lambda}{\partial t}$$

$$= -v\frac{\partial u_1}{\partial z} = -v\frac{dg_1}{d\lambda}\frac{\partial \lambda}{\partial z}$$

Since

$$\frac{\partial \lambda}{\partial t} = 1; \quad \frac{\partial \lambda}{\partial z} = -1/v$$

substitution in the preceding equation gives

$$\frac{\partial u_1}{\partial t} = \frac{dg_1}{d\lambda}(1)$$

$$= -v\frac{\partial u_1}{\partial z} = -v\frac{dg_1}{d\lambda}(-1/v) \ (QED^{11})$$

so that eq. (1.6a) is a solution to eq. (1.5a).

Also, eq. (1.6a) satisfies IC (1.4a) (with $f_1(z) = 0$) and BC (1.4b) (with $z = 0$). Therefore, eq. (1.6a) is the analytical solution to eqs. (1.5a) to (1.5c). It is termed a *traveling wave solution* [1] since it depends only on the *Lagrangian* variable $\lambda = t - z/v$.

For the special case of $g_1(t) = h(t)$ of eq. (1.5d), the analytical solution is

$$u_1(z,t) = h(t - z/v) \qquad (1.6b)$$

---

[11]QED ="quod erat demonstrandu" (Latin) or "that which was to be demonstrated".

Eq. (1.6b) defines a unit step[12], starting at $z = 0$ and traveling in the direction of increasing z (up the chromatographic column of Fig. 1.1) with velocity v. The unit step occurs at $\lambda = t - z/v = 0$. Thus, at the exit of the column, $\lambda = t - z_L/v = 0$ or at $t = z_L/v = 50/1 = 50$). This unit step is evident in Fig. 1.2, except that the step is actually undefined (not single valued) and is approximated over three grid points as programmed in step of Listing 1.3.

The fact that $u_1(z = z_L, t)$ is not a discontinuity at $t = 50$ is clear in Fig. 1.2 (the slope is finite). However, the approximation is required since a function that is not single valued (and with an infinite slope) cannot be programmed (it can be plotted using two points, $u_1(\lambda = 0) = 0$ and $u_1(\lambda = 0) = 1$ (both values used at $t = 50$), but this is an artifice just to give the step the required appearance).

As another perspective, eqs. (1.5) constitute an impossible problem numerically since the derivative (slope) $\dfrac{\partial u_1}{\partial z}$ at $t - z/v = 0$ is infinite. For the purpose of computing a numerical solution, this discontinuity is approximated by a function with a finite slope as in function step of Listing 1.3. In the subsequent discussion, we will consider how closely the methods for calculating $\dfrac{\partial u_1}{\partial z}$ produce the solution of eq. (1.6b), that is, for ifd = 1,2,3,4 in Listing 1.1.

In summary, eq. (1.5a) is an elementary hyperbolic PDE for which an analytical solution is easily derived, yet it is one of the most difficult PDEs to integrate numerically (since it propagates steep fronts and discontinuities, a general feature of hyperbolic PDEs).

- The numerical solution of eqs. (1.1) to (1.4) (in Fig. 1.2) plotted with o is a smoothed (rounded) approximation of the unit step solution of eq. (1.6b). This smoothing is generally termed *numerical diffusion* and is one of two distortions (numerical artifacts) of solutions with steep moving fronts or discontinuities. The other distortion is *numerical oscillation* that is described next. Numerical diffusion may preclude the accurate calculation of moving front solutions in applications for which this is unacceptable, e.g., chromatography, as discussed subsequently.

- The two-point upwind FD approximations used in the numerical solution of eqs. (1.1) to (1.4) are illustrated by the following code taken from dss012 (ifd=1) listed in Appendix B.

```
for(i in 2:n){
    ux[i]=(u[i]-u[i-1])/dx;
}
```

with n=41 in pde_1 of Listing 1.2 where dss012 is called. The derivative, ux[i], is approximated at grid point i by a FD based on a weighted sum of u[i] and u[i-1] with a spacing dx (for v > 0). Point i-1 is upwind (upstream) of point i. This upwinding is essential in the case of steep moving fronts and discontinuities in the

---

[12] A PDE with a solution that depends only on an IC is generally termed a *Cauchy problem*, that is, an initial value problem. If the IC is discontinuous, the PDE is termed a *Riemann problem*. Eqs. (1.5) are an example of a Riemann problem since they define a PDE problem with a solution that has a discontinuity for $t \geq 0$, that is, a discontinuity at $t - z/v = 0$ as stated in eq. (1.6b).

solution. If u[i+1] is used in place of u[i-1], the numerical solution will become unstable. Physically, this makes sense since what happens at i is determined by what is happening upstream at i-1 and not downstream at i+1.

- This use of upwinding requires a priori knowledge of the direction of flow, e.g., bottom to top in the chromatographic column of Fig. 1.1. An incorrect direction used in the FD approximation leads to unstable solutions. For example, if v < 0, the preceding FD approximation will produce an unstable numerical solution. Rather, u[i] and u[i+1] are used in the FD since i+1 is now in the upstream direction (see the listing of dss012 in Appendix B). Approximations that are *centered* (rather than upwinded), and therefore do not require knowledge of the direction of flow, are discussed subsequently (for ifd=2).

- For i=2 (one point inside the left boundary in z), u[1] is required (in the preceding code). u[1] is set as a BC, as illustrated in pde_1 of Listing 1.2.

In summary, two-point upwinding generally produces a stable solution with no numerical oscillation, but with numerical diffusion that may be excessive, depending on the application, such as for hyperbolic (convective) PDEs that propagate steep fronts and discontinuities.

For ncase=2 in Listing 1.1, Fig. 1.3 results (the numerical output is in Table 1.2). We can note the following details in Fig. 1.3.

- The step input, $u_1(z = 0, t) = h(t)$, is smoothed beyond that from the two-point upwind approximation of Fig. 1.2 by the adsorption onto the adsorbate. Whether this is an accurate solution is difficult to assess without some form of *error analysis*, for example, h and p refinement, which are considered next. As is generally the case, an analytical solution is not readily available that can be used to calculate an exact error as in Fig. 1.2. A principal reason for not having an analytical solution is the nonlinearity of the rate, that is, the term $u_1(u_1^e - u_2)$ with the $u_1 u_2$ product, and the isotherm of eq. (1.3) which is nonlinear in u_2 for $c_2 \neq 0$. In other words, an analytical solution is precluded, and we accept a numerical solution that we expect will be of reasonable accuracy, e.g., 3-4 significant figures. However, this expectation must be justified (as it should for all PDE numerical solutions).

- An important consideration in evaluating the numerical solution of Table 1.2 and Fig. 1.3 is whether the numerical diffusion from the two-point upwind approximation is significant relative to the physical smoothing from the adsorption. This point requires further investigation through h and p refinement which is facilitated by having the choice of four approximations selected with ifd. To start, we consider the numerical solution of eqs. (1.1) to (1.4) with ifd=2 in the main program of Listing (1.1) (everything else remains the same).

idf = 2 gives a call to dss004 in pde_1. dss004 has five-point centered (fourth-order correct) FD approximations (except near the boundaries ($z = 0, z_L$) where noncentered FDs are used) so we would expect improved accuracy of the solution relative to ifd=1 (five points rather than two). However, we will observe this is incorrect.

The numerical output for ifd=2 is given in Table 1.3

```
ifd =  2    ncase =  1
```

t	u1(z=zL,t)	rate(z=zL,t)
0.00	0.0000	0.0000
3.00	0.0000	0.0000
.	.	
.	.	
.	.	

(output for t = 6 to 36 deleted)

.	.	
.	.	
.	.	
39.00	0.0003	0.0000
42.00	-0.0019	0.0000
45.00	-0.0149	0.0000
48.00	0.0850	0.0000
51.00	0.6330	0.0000
54.00	1.1941	0.0000
57.00	0.9481	0.0000
60.00	0.9382	0.0000
63.00	1.1158	0.0000
66.00	0.8752	0.0000
69.00	1.1072	0.0000
72.00	0.9229	0.0000
75.00	1.0437	0.0000
78.00	0.9880	0.0000
81.00	0.9843	0.0000
84.00	1.0384	0.0000
87.00	0.9439	0.0000
90.00	1.0693	0.0000
93.00	0.9215	0.0000
96.00	1.0844	0.0000
99.00	0.9125	0.0000
102.00	1.0885	0.0000
105.00	0.9122	0.0000
108.00	1.0858	0.0000
111.00	0.9171	0.0000
114.00	1.0794	0.0000
117.00	0.9248	0.0000
120.00	1.0710	0.0000
123.00	0.9321	0.0000
126.00	1.0620	0.0000
129.00	0.9456	0.0000
132.00	1.0548	0.0000
135.00	0.9422	0.0000

```
138.00         1.0490          0.0000
141.00         0.9667          0.0000
144.00         1.0267          0.0000
147.00         0.9660          0.0000
150.00         1.0459          0.0000


ncall = 1311



ifd =  2    ncase =  2

     t      u1(z=zL,t)  rate(z=zL,t)
   0.00       0.0000        0.0000
   3.00       0.0000        0.0000

              .             .
              .             .
              .             .
   (output for t = 6 to 36 deleted)
              .             .
              .             .
              .             .
  39.00       0.0001        0.0000
  42.00      -0.0005        0.0000
  45.00      -0.0049        0.0000
  48.00       0.0036        0.0000
  51.00       0.0337        0.0005
  54.00       0.0654        0.0010
  57.00       0.1012        0.0017
  60.00       0.1460        0.0029
  63.00       0.2051        0.0048
  66.00       0.2886        0.0081
  69.00       0.4133        0.0136
  72.00       0.5950        0.0200
  75.00       0.8073        0.0195
  78.00       0.9486        0.0089
  81.00       0.9866        0.0015
  84.00       0.9945        0.0007
  87.00       1.0004        0.0003
  90.00       0.9996       -0.0001
  93.00       0.9996        0.0001
  96.00       1.0004        0.0000
  99.00       0.9998       -0.0000
 102.00       0.9999        0.0000
 105.00       1.0001       -0.0000
 108.00       0.9999       -0.0000
```

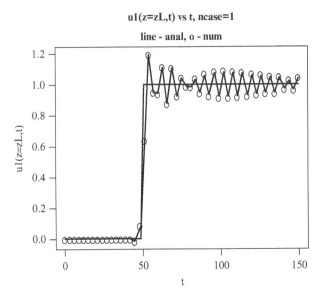

Figure 1.4   Comparison of the numerical and analytical solutions of eqs. (1.1b), (1.2b)

```
111.00        1.0000        0.0000
               .              .
               .              .
               .              .
(output for t = 114 to 144 deleted)
               .              .
               .              .
               .              .
147.00        1.0000       -0.0000
150.00        1.0000       -0.0000

ncall = 1035
```

Table 1.3: Numerical output for eqs. (1.1) to (1.4) for `ncase=1,2`, `ifd=2`

Figs. 1.4, 1.5 follow.
We can note the following points from Table 1.3 and Figs. 1.4, 1.5.

- In Table 1.3, `ncase=1`, and in Fig. 1.4, the numerical solution is highly oscillatory (termed *numerical oscillation*). This is the second form of numerical distortion (in addition to numerical diffusion illustrated in Fig. 1.2). Clearly, by comparison with the analytical solution, the numerical solution is unacceptable (and it is also unrealistic physically since we would not expect the output from the adsorption column of Fig. 1.1 to oscillate).

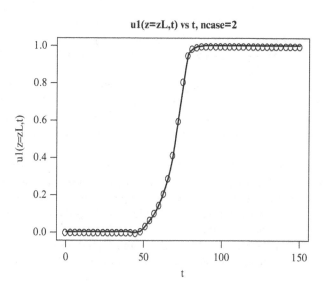

Figure 1.5 Numerical solution of eqs. (1.1b), (1.2b) for `ncase=2`, `ifd=2`

- In Fig. 1.5, the numerical oscillation is essentially eliminated by the adsorption (`ncase=2`), which demonstrates that the performance of a spatial differentiator may be strongly dependent on the characteristics (features) of the PDE problem. For example, with no adsorption (`ncase=1`) (so that the solution is the propagation of a unit step or discontinuity), the two-point upwind (2pu) approximation gives excessive numerical diffusion (Fig. 1.2) and the five-point centered approximation (5pc) gives numerical oscillation (Fig. 1.4). However, for the 2pu with adsorption (`ncase=2`), the diffusion still appears to be unacceptable (Fig. 1.3) while for 5pc, the physical smoothing was sufficient to give what appears to be an accurate solution (Fig. 1.5). But this use of p refinement (2pu to 5pc) is inconclusive and further analysis is required.

The general conclusion we reach is that centered FD approximations should not be used for strongly hyperbolic (convective) PDEs with a solution that includes a steep moving front or discontinuity. This conclusion remains valid if the number of grid points is increased (e.g., n=41 to 81). In fact, the oscillations usually become even more pronounced[13]. This conclusion also remains valid if the order of the FD approximation is increased, e.g., seven-point centered approximations oscillate as much as the five-point FD approximations.

To complete this discussion of five-point centered approximations, a section of code from `dss004` (listed in Appendix B) is given below.

---

[13]We should not conclude that five-point centered FDs are always unsatisfactory. In fact, they generally work very well for parabolic (diffusive) PDEs such as the *heat conduction equation* (*Fourier's second law*) and the *diffusion equation* (*Fick's second law*). This point will be illustrated in later applications.

```
#
# Interior points (x=xl+2*dx,...,x=xu-2*dx)
  for(i in 3:(n-2))ux[i]=r12dx*(-u[i+2]+8*u[i+1]-8*u[i-1]+u[i-2]);
```

Note that the derivative ux[i] is computed as a weighted sum of five values of u with weighting coefficients -1 8 0 -8 1 that are skew symmetric with respect to the center point i where the coefficient is 0 (and is therefore not programmed). At the boundary points i=1,2,n-1,n, noncentered approximations are used. Details are given in dss004 listed in Appendix B.

The question then naturally arises if there is a FD approximation with acceptable levels of numerical diffusion and oscillation. If the two-point upwind approximations (in dss012) do not oscillate and the five-point approximations (in dss004) have a relatively low level of numerical diffusion (see Fig. 1.4 along the near vertical analytical solution) perhaps somehow combining the approximations would be worth trying. To this end, we consider the five-point biased upwind (5pbu) approximations in dss020 listed in Appendix B. A section of the coding from dss020 for the derivative at i is listed below.

```
for(i in 4:(n-1)){
  ux[ i]=r12dx*(  -u[i-3] +6*u[i-2]-18*u[i-1]+10*u[i ]+3*u[i+1]); }
```

The derivative ux[i] is the weighted sum of five values of u with grid indices i-3,i-2,i-1,i,i+1 (for v > 0). That is, three points upstream of i and one point downstream are used. This biasing in the upstream direction, designated as 5pbu, is intended to maintain the effect of the flow. The effectiveness of this approach is reflected in the numerical output in Table 1.4 and Figs. 1.6 and 1.7 (produced with ifd=3 in main program of Listing 1.1).

```
 ifd =  3    ncase =  1

     t      u1(z=zL,t)  rate(z=zL,t)
   0.00       0.0000       0.0000
   3.00       0.0000       0.0000
                  .            .
                  .            .
                  .            .
   (output for t = 6 to 24 deleted)
                  .            .
                  .            .
                  .            .
  27.00       0.0001       0.0000
  30.00      -0.0005       0.0000
  33.00       0.0015       0.0000
  36.00      -0.0019       0.0000
  39.00      -0.0074       0.0000
  42.00       0.0522       0.0000
```

```
45.00     -0.1178      0.0000
48.00      0.1066      0.0000
51.00      0.7831      0.0000
54.00      1.0120      0.0000
57.00      1.0039      0.0000
60.00      1.0026      0.0000
63.00      0.9980      0.0000
66.00      1.0012      0.0000
69.00      0.9994      0.0000
72.00      1.0003      0.0000
75.00      0.9999      0.0000
78.00      1.0000      0.0000
             .            .
             .            .
             .            .
```

(output for t = 81 to 144 deleted)

```
             .            .
             .            .
             .            .
147.00      1.0000      0.0000
150.00      1.0000      0.0000

ncall =   914


ifd =   3   ncase =   2

    t      u1(z=zL,t)  rate(z=zL,t)
  0.00      0.0000       0.0000
  3.00      0.0000       0.0000
             .            .
             .            .
             .            .
```

(output for t = 6 to 24 deleted)

```
             .            .
             .            .
             .            .
27.00      0.0001       0.0000
30.00     -0.0002       0.0000
33.00      0.0005       0.0000
36.00     -0.0005      -0.0000
39.00     -0.0020      -0.0000
42.00      0.0081       0.0000
45.00     -0.0086       0.0000
48.00      0.0012      -0.0000
51.00      0.0304       0.0004
```

```
54.00          0.0606          0.0009
57.00          0.0958          0.0016
60.00          0.1400          0.0028
63.00          0.1998          0.0048
66.00          0.2879          0.0086
69.00          0.4238          0.0149
72.00          0.6143          0.0203
75.00          0.8101          0.0171
78.00          0.9366          0.0086
81.00          0.9865          0.0027
84.00          0.9987          0.0005
87.00          1.0002          0.0000
90.00          1.0001         -0.0000
93.00          1.0000         -0.0000

                 .               .
                 .               .
                 .               .

(output for t = 96 to 144 deleted)

                 .               .
                 .               .
                 .               .

147.00         1.0000         -0.0000
150.00         1.0000         -0.0000


ncall = 1312
```

Table 1.4: Numerical output for eqs. (1.1) to (1.4) for ncase=1,2, ifd=3

Figs. 1.6, 1.7 follow.
We can note the following points from Table 1.4 and Figs. 1.6, 1.7.

- In Fig. 1.6, the 5pbu approximation has substantially reduced the numerical diffusion and oscillation of Figs. 1.2 and 1.4. However, there is an oscillation at the leading edge of the numerical solution that could still render the numerical solution unacceptable. But we should keep in mind that the numerical solution approximates a unit step, which is essentially an impossible requirement (as discussed previously), so that the numerical solution is a substantial improvement over the previous 2pu (ifd=1) and 5pc (ifd=2) solutions.

- In Fig. 1.7, the oscillation of Fig. 1.6 has been essentially eliminated by the adsorption (for ncase=2) as occurred in Fig. 1.5. Also, Figs. 1.5 and 1.7 are quite similar which suggests that they reflect an accurate solution (although this is certainly not a proof of accuracy and some additional cases with a number of grid points other than n=41 should be considered).

To demonstrate this point of the similarity of the solutions for ifd=2,3, from Tables 1.3 and 1.4, we have at the portions of the solutions changing most rapidly (see Figs. 1.5, 1.7):

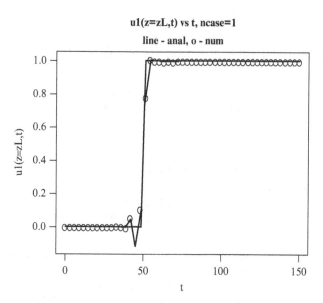

Figure 1.6 Comparison of the numerical and analytical solutions of eqs. (1.1b), (1.2b)

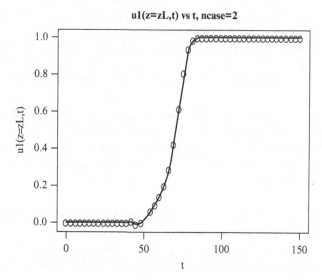

Figure 1.7 Numerical solution of eqs. (1.1b), (1.2b), `ncase=2`, `ifd=3`

```
t = 63
  5pc  (ifd=2)
  63.00        0.2051        0.0048
  5pbu (ifd=3)
  63.00        0.1998        0.0048 (n=41)
  63.00        0.2026        0.0048 (n=81)
```

```
t = 69
  5pc  (ifd=2)
  69.00        0.4133        0.0136
  5pbu (ifd=3)
  69.00        0.4238        0.0149 (n=41)
  69.00        0.4167        0.0142 (n=81)

t = 75
  5pc  (ifd=2)
  75.00        0.8073        0.0195
  5pbu (ifd=3)
  75.00        0.8101        0.0171 (n=41)
  75.00        0.8157        0.0182 (n=81)
```

For 5pbu, solutions are summarized with n=41, 81 to demonstrate the level of convergence from h refinement.

Experience has indicated that the 5pbu frequently works as required to produce an accurate numerical solution if the moving front of the solution is not very steep, which is the case in many physical applications. But the preceding results (for ifd=1,2,3) suggest that some experimentation and careful evaluation of the numerical solution is usually required, including the use of h refinement, that is, changing the number of grid points and observing the effect on the numerical solution. We have not done that here because of space limitations, but rather used only n=41 (with results for n=81 indicated in the preceding table). Changes in the number of grid points requires only changing n in the main program of Listing 1.1.

We conclude this discussion of FD approximation of hyperbolic PDEs that propagate steep fronts and discontinuities with Godunov's barrier theorem that pertains to numerical diffusion and oscillation. This theorem states ([1], p25): There is no linear approximation to the Riemann problem, higher than first order, that is nonoscillatory.

To explain the wording:

- ncase=1 in the previous examples corresponds to the Riemann problem (the unit step or discontinuity BC of eqs. (1.4b) and (1.5b)).
- The FD approximations 2pu, 5pc and 5pbu are linear in the sense that u[i] in the RHS weighted sums for the calculation of ux[i] is to the first power (see the preceding portions of code).
- 2pu is first order and does not oscillate (Fig. 1.2).
- 5pc and 5pbu are fourth order (i.e., higher than first order) and oscillate (Figs. 1.4, 1.6).

Thus, if we are to use a higher order method to achieve better accuracy (e.g., less diffusion than 2pu), we will have to use a nonlinear approximation or algorithm to avoid oscillation. This is the approach considered next based on *flux limiters*.

(1.3.2) Flux limiters, step BC

Flux limiters provide a nonlinear approximation to the first-order spatial derivatives in convective (hyperbolic) PDEs, e.g., $\dfrac{\partial u_1}{\partial z}$ in eq. (1.1b). The nonlinear algorithm can be used to eliminate numerical oscillations as explained by Godunov's theorem cited previously.

For example, we can code the van Leer flux limiter ([1], pp 37-43) in the format of the 5pc and 5pbu FD approximations. This has been done in function vanl listed in Appendix B. We can then call vanl by using ifd=4 in the main program of Listing 1.1. Abbreviated numerical output is listed in Table 1.5 below, and the graphical output is in Figs. 1.8 and 1.9.

```
ifd =  4    ncase =  1
    t        u1(z=zL,t)   rate(z=zL,t)
  0.00        0.0000       0.0000
  3.00       -0.0000       0.0000
                .             .
                .             .
                .             .

(output for t = 6 to 42 deleted)
                .             .
                .             .
                .             .

 45.00        0.0022       0.0000
 48.00        0.2420       0.0000
 51.00        0.6735       0.0000
 54.00        0.9253       0.0000
 57.00        0.9899       0.0000
 60.00        0.9989       0.0000
 63.00        0.9999       0.0000
 66.00        1.0000       0.0000
                .             .
                .             .
                .             .

(output for t = 69 to 144 deleted)
                .             .
                .             .
                .             .

147.00        1.0000       0.0000
150.00        1.0000       0.0000

ncall = 2191


ifd =  4    ncase =  2
```

```
   t      u1(z=zL,t)   rate(z=zL,t)
  0.00      0.0000       0.0000
  3.00     -0.0000       0.0000
               .             .
               .             .
               .             .
  (output for t = 6 to 45 deleted)
               .             .
               .             .
               .             .
 48.00      0.0063       0.0000
 51.00      0.0334       0.0004
 54.00      0.0666       0.0011
 57.00      0.1055       0.0020
 60.00      0.1549       0.0033
 63.00      0.2223       0.0057
 66.00      0.3180       0.0096
 69.00      0.4511       0.0143
 72.00      0.6151       0.0169
 75.00      0.7770       0.0146
 78.00      0.8969       0.0091
 81.00      0.9621       0.0042
 84.00      0.9883       0.0015
 87.00      0.9968       0.0005
 90.00      0.9992       0.0001
 93.00      0.9998       0.0000
 96.00      0.9999       0.0000
 99.00      1.0000       0.0000
               .             .
               .             .
               .             .
  (output for t = 102 to 144 deleted)
               .             .
               .             .
               .             .
147.00      1.0000       0.0000
150.00      1.0000       0.0000

ncall = 5620
```

Table 1.5: Numerical output for eqs. (1.1) to (1.4) for ncase=1,2, ifd=4

Fig. 1.8 indicates that the unit step of eq. (1.5d) is closely approximated, with little numerical diffusion and no oscillation. The latter is termed *essentially non-oscillatory* or ENO.

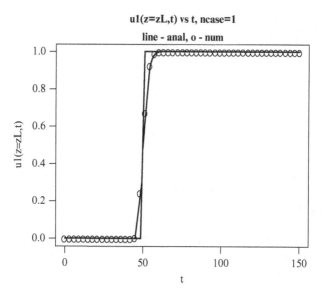

Figure 1.8 Comparison of the numerical and analytical solutions of eqs. (1.1b), (1.2b)

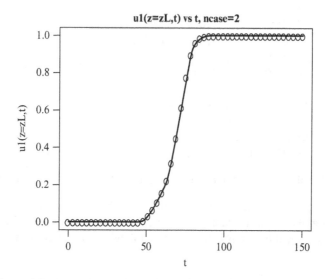

Figure 1.9 Numerical solution of eqs. (1.1b), (1.2b), ncase=2, ifd=4

Fig. 1.9 also indicates a smooth solution (no oscillation), but this was also achieved with 5pc and 5pbu (Figs. 1.5, 1.7) because of smoothing of the transfer to the adsorbent for ncase=2.

The three approaches, 2pc, 5pbu and van Leer, are briefly compared in Table 1.6

```
t = 63
  5pc  (ifd=2)
  63.00        0.2051        0.0048
  5pbu (ifd=3)
  63.00        0.1998        0.0048 (n=41)
  63.00        0.2026        0.0048 (n=81)
  van Leer (ifd=4)
  63.00        0.2223        0.0057

t = 69
  5pc  (ifd=2)
  69.00        0.4133        0.0136
  5pbu (ifd=3)
  69.00        0.4238        0.0149 (n=41)
  69.00        0.4167        0.0142 (n=81)
  van Leer (ifd=4)
  69.00        0.4511        0.0143

t = 75
  5pc  (ifd=2)
  75.00        0.8073        0.0195
  5pbu (ifd=3)
  75.00        0.8101        0.0171 (n=41)
  75.00        0.8157        0.0182 (n=81)
  van Leer (ifd=4)
  75.00        0.7770        0.0146

  Calls to pde_1, ncase=2
    5pc        ncall = 1035
    5pbu       ncall = 1312
    van Leer   ncall = 5620
```

Table 1.6: Abbreviated comparison of output for eqs. (1.1) to (1.4) for ncase=2

The differences in the numerical solutions in Table 1.6 (for ifd=2,3,4) suggest that these differences are substantial. However, this is not necessarily the case as indicated by the graphical output produced by the following variant of the main program of Listing 1.1.

```
#
# Delete previous workspaces
  rm(list=ls(all=TRUE))
#
#   1D, one component, chromatography model
#
#   The ODE/PDE system is
#
```

```
#    u1_t = -v*u1_z - (1 - eps)/eps*rate                        (1.1b)
#
#    u2_t = rate                                                (1.2b)
#
#    rate = kf*u1*(u2eq - u2) - kr*u2
#
#    u2eq = c1*u1/(1 + c2*u1)                                   (1.3)
#
#    Boundary condition
#
#      u1(z=0,t) = step(t)                                      (1.4b)
#
#    Initial conditions
#
#      u1(z,t=0) = 0                                            (1.4a)
#
#      u2(z,t=0) = 0                                            (1.4c)
#
#    The method of lines (MOL) solution for eqs. (1.1) to
#    (1.4) is coded below.  Specifically, the spatial
#    derivative in the fluid balance, u1_z in eq. (1.1b),
#    is replaced by one of four approximations as selected
#    by the variable ifd.
#
# Access ODE integrator
  library("deSolve");
#
# Access files
  setwd("G:/comp3/chromatography/R/ex1");
  source("pde_1.R") ;source("step.R")   ;
  source("dss004.R");source("dss012.R");
  source("dss020.R");source("vanl.R")   ;
#
# Declare (preallocate) array for plotted solutions
  nout=51;
  plot_2=matrix(0,nrow=nout,ncol=2);
#
# Step through cases
  for(ncase in 1:2){
#
#   Model parameters
      v=1;  eps=0.4;  u10=0;  u20=0;
    c1=1;     c2=1;  kf=1;   kr=1;
    zL=50;    n=41;
    if(ncase==1){ ifd=3; }
    if(ncase==2){ ifd=4; }
```

```
#
# Level of output
#
#   Detailed output   - ip = 1
#
#   Brief (IC) output - ip = 2
#
  ip=2;
#
# Initial condition
  u0=rep(0,2*n);
  for(i in 1:n){
      u0[i]=u10;
    u0[i+n]=u20;
  }
  t0=0;tf=150;
  tout=seq(from=t0,to=tf,by=(tf-t0)/(nout-1));
  ncall=0;
#
# ODE integration
  out=ode(func=pde_1,times=tout,y=u0);
#
# Store solution
  u1=matrix(0,nrow=nout,ncol=n);
  u2=matrix(0,nrow=nout,ncol=n);
  t=rep(0,nout);
  for(it in 1:nout){
  for(iz in 1:n){
    u1[it,iz]=out[it,iz+1];
    u2[it,iz]=out[it,iz+1+n];
  }
    t[it]=out[it,1];
  }
#
# Display ifd, ncase
  cat(sprintf("\n ifd = %2d   ncase = %2d",ifd,ncase));
#
# Display numerical solution
  if(ip==1){
  cat(sprintf(
    "\n\n     t      u1(z=zL,t)  rate(z=zL,t)\n"));
  u2eq=rep(0,nout);rate=rep(0,nout);
  for(it in 1:nout){
    u2eq[it]=c1*u1[it,n]/(1+c2*u1[it,n]);
    rate[it]=kf*u1[it,n]*(u2eq[it]-u2[it,n])-kr*u2[it,n];
```

```
    cat(sprintf(
      "%7.2f%12.4f%12.4f\n",t[it],u1[it,n],rate[it]));
  }
  }
#
# Store solution for plotting
  tplot=rep(0,nout);
  for(it in 1:nout){
    plot_2[it,ncase]=u1[it,n];
    tplot[it]=t[it];
  }
#
# Calls to ODE routine
  cat(sprintf("\n ncall = %4d\n",ncall));
#
# Next case
  }
#
# Plot for u1(z=zL,t)
  par(mfrow=c(1,1))
  plot(tplot,plot_2[,1],
      xlab="t",ylab="u1(z=zL,t)",
      xlim=c(0,tplot[nout]),ylim=c(0,1),
      main="1 - 5pbu, 2 - van Leer",
      type="l",lwd=2);
   points(tplot,plot_2[,1], pch="1",lwd=2);
    lines(tplot,plot_2[,2],type="l",lwd=2);
   points(tplot,plot_2[,2], pch="2",lwd=2);
```

Listing 1.4: Main program pde_1_main for comparison of the 5pbu and van Leer solutions for ncase=2

Listing 1.4 is similar to Listing 1.1, so we note only the differences here.

- The routine vanl for the van Leer flux limiter is included.

    ```
    source("dss020.R");source("vanl.R");
    ```

- An array (matrix) is defined with the matrix utility for the two solutions ifd=3,4, ncase=2.

    ```
    #
    # Declare (preallocate) array for plotted solutions
      nout=51;
      plot_2=matrix(0,nrow=nout,ncol=2);
    ```

In this way, the two solutions can be superimposed on the same plot (Figs. 1.10, 1.11).

- Two cases are programmed corresponding to `ifd=3,4` (5pbu, van Leer).

```
#
# Step through cases
  for(ncase in 1:2){
#
#    Model parameters
        v=1; eps=0.4; u10=0; u20=0;
      c1=1;     c2=1;    kf=1;   kr=1;
      zL=50;      n=41;
      if(ncase==1){ ifd=3; }
      if(ncase==2){ ifd=4; }
```

- Brief numerical output is selected.

```
  ip=2;
```

- The numerical solution for `ncase=1,2` (`ifd=3,4`) is placed in array `plot_2`.

```
#
# Store solution for plotting
  tplot=rep(0,nout);
  for(it in 1:nout){
    plot_2[it,ncase]=u1[it,n];
    tplot[it]=t[it];
  }
```

Note that this is at `i=n` corresponding to $z = z_L$.

- At the end of the second solution, both solutions (for 5pbu and van Leer) are plotted as a composite plot identified with 1 and 2.

```
#
# Plot for u1(z=zL,t)
  par(mfrow=c(1,1))
  plot(tplot,plot_2[,1],
      xlab="t",ylab="u1(z=zL,t)",
      xlim=c(0,tplot[nout]),ylim=c(0,1),
      main="1 - 5pbu, 2 - van Leer",
      type="l",lwd=2);
  points(tplot,plot_2[,1], pch="1",lwd=2);
    lines(tplot,plot_2[,2],type="l",lwd=2);
  points(tplot,plot_2[,2], pch="2",lwd=2);
```

The composite plot is produced with a combination of three utilities, `plot`, `lines`, `points`. The result is in Fig. 1.10, and when the two `points` are not included (by making those statements comments), Fig. 1.11 results.

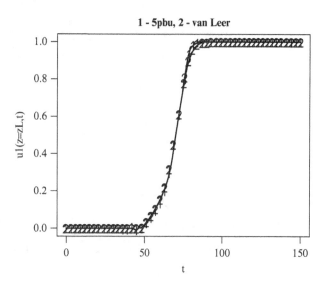

Figure 1.10   Comparison of 5pbu (`ifd=3`) and van Leer (`ifd=4`)

The numerical output is

```
ifd =  3   ncase =  1
ncall = 1312

ifd =  4   ncase =  2
ncall = 5620
```

so that the van Leer limiter (from `idf=4`) requires substantially more computation than 5pbu (from `ifd=3`).

Fig 1.10 indicates that the two solutions agree closely. This is further confirmed in Fig. 1.11 in which the numbered points have been surppressed.

The fact that the two solutions agree closely does not prove that they are accurate and correct. However, this agreement resulting from two different algorithms, 5pbu and van Leer, suggests that the two solutions are accurate. We can view this approach of comparing solutions from two different algorithms as a generalized form of p refinement in which not only is the order of the approximation changed (p usually denotes the order), but the algorithm itself is changed.

To study this approach, we could consider other flux limiters. A set of limiters is provided in [1], pp 40-43, and these can easily be used in place of the van Leer limiter in `vanl`.

This completes the discussion of the model of eqs. (1.1) to (1.4). In particular, the unit step of eq. (1.5d) provides a stringent test of the numerical algorithms (within the MOL format). We now go through a similar analysis using a less stringent BC function, a pulse in place of the unit step.

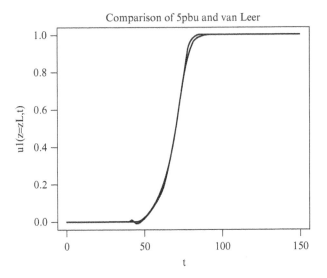

Figure 1.11   Comparison of 5pbu (`ifd=3`) and van Leer (`ifd=4`)

(1.3.3) FDs, pulse BC

$g_1(t)$ in BC (1.4b) is a cosine pulse, defined as a function of the Lagrangian variable $(t - z/v)$.

$$\text{pulse(t)} = \begin{cases} 0, & (t - z/v) < 0 \\ 1 - \cos(\omega(t - z/v)), & 0 \le (t - z/v) < \pi/2 \\ 1 + \cos(\omega(t - z/v)), & \pi/2 \le (t - z/v) \le \pi \\ 0, & (t - z/v) > \pi \end{cases} \tag{1.7a}$$

pulse(t) is a smooth (continuous) function of t in contrast with the step of eq. (1.5d). Therefore, we would expect that calculating solutions to eqs. (1.1) to (1.4) would be easier than for the step function. However, it is included in this analysis since for the multi component case considered subsequently, we can observe the separation of the component pulses as would occur in a chromatographic column.

The pulse of eq. (1.7a) is programmed in function `pulse`.

```
pulse=function(t,z,v) {
#
# Function pulse computes a pulse function
#
  w=0.05;tzv=t-z/v;wtzv=w*tzv;
  if((wtzv)< 0 ){u1p=0;}
  if((wtzv)>=0    )&(wtzv< pi/2)) {u1p=1-cos(wtzv);}
  if((wtzv)>=pi/2)&(wtzv<=pi  )) {u1p=1+cos(wtzv);}
  if((wtzv)>pi) {u1p=0;}
#
```

```
# Return pulse
  return(c(u1p));
}
```

Listing 1.5: Function `pulse` from eq. (1.7a)

The programming in Listing 1.5 follows directly from eq. (1.7a) and therefore does not require elaboration. Note that $\omega = 0.05$ which was selected to give a pulse with a suitable spread for the graphical output (plots of the following figures) for $0 \le t \le 150$.

pulse of Listing 1.5 is used in Listings 1.1 to 1.3 by merely replacing the use of step with pulse. So discussion of the programming details is not required. The exact solution for ncase=1 is, from eq. (1.6a)

$$u_1(z, t) = \text{pulse}(t - z/v) \tag{1.7b}$$

As before, the analytical solution of eq. (1.7b) can be used to give the exact error in the numerical solution for ncase=1 for the various spatial differentiation routines (ifd=1,2,3,4). We would expect that the agreement between the numerical and analytical solutions would be better than for the step since the pulse of eq. (1.7a) is smoother than the step of eq. (1.5d).

The numerical and graphical output from Listings 1.1 to 1.3 follows.

In Fig. 1.12 (ncase=1), the numerical diffusion for 2pu (ifd=1) is substantial, particularly at the peak which is reduced from 1 to approximately 0.7. The defined vertical scaling was used in producing this plot (ylim=c(0,1)) since the numerical solution is

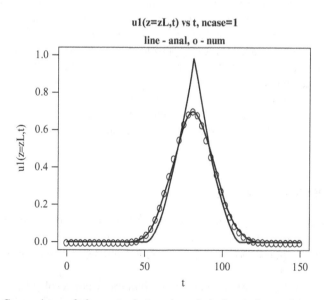

Figure 1.12 Comparison of the numerical and analytical solutions of eqs. (1.1b), (1.2b) ncase=1, ifd=1, pulse BC

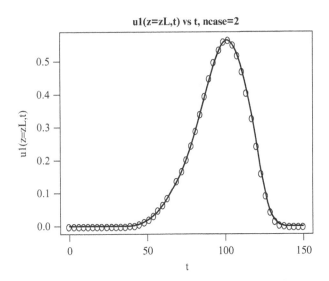

Figure 1.13   Numerical solution of eqs. (1.1b), (1.2b), ncase=2, ifd=1, pulse BC

plotted first. Automatic scaling gives a vertical axis of 0 to 0.8, and then the peak of the exact solution plotted next is truncated (the plot cannot reach 1).

In Fig. 1.13 (ncase=2), the numerical solution appears smooth, but the accuracy of the solution cannot be ascertained. Later we compare the ncase=2 solutions for different spatial differentiators.

Execution of the routines for ifd=2 gives the graphical output in Fig. 1.14 (the numerical output and the plotted solution for ncase=2 are not given here).

In Fig. 1.14, the numerical and analytical solutions agree closely, except for some numerical oscillation in the downstream portion of the solution, which is not unexpected since we found previously that the 5pc approximations (ifd=2) oscillate (for steep moving fronts such as the unit step).

Execution of the routines for ifd=3 gives the graphical output in Fig. 1.15 (the numerical output and the plotted solution for ncase=2 are not given here).

In Fig. 1.15, the numerical and analytical solutions agree closely, with no apparent numerical diffusion or oscillation.

(1.3.4) Flux limiters, pulse BC

We next consider the solutions to eqs. (1.1b) and (1.2b) with the derivative $\dfrac{\partial u_1}{\partial z}$ in eq. (1.1b) approximated with a flux limiter. Execution of the routines for ifd=4 gives the graphical output in Fig. 1.16 (the numerical output and the plotted solution for ncase=2 are not given here).

In Fig. 1.16, the peak of the numerical solution is not resolved as closely as we might expect. Also, the van Leer limiter (ifd=4) required substantially more calculations (higher value of ncall) than the 5pbu FD (ifd=3).

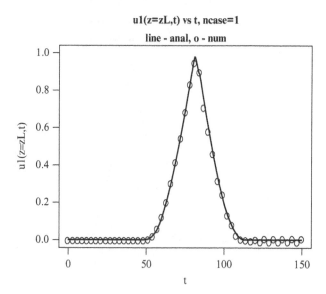

Figure 1.14 Comparison of the numerical and analytical solutions of eqs. (1.1b), (1.2b) `ncase=1`, `ifd=2`, pulse BC

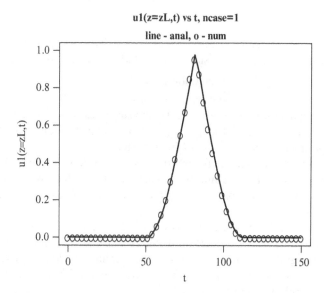

Figure 1.15 Comparison of the numerical and analytical solutions of eqs. (1.1b), (1.2b) `ncase=1`, `ifd=3`, pulse BC

The error in the numerical solution at the peak is not unexpected when we consider how rapidly the solution changes at the peak. In fact, the first derivative of the solution is discontinuous at the peak. To show this, the derivative from the segment $u_1 = 1 - \cos(\omega(t - z/v))$ (from eq. (1.7a)) is $\omega \sin(\omega(t - z/v))$ and at the peak, $\omega(t - z/v) = \pi/2$ the derivative is ω. The derivative from the segment $u_1 = 1 + \cos(\omega(t - z/v))$ is

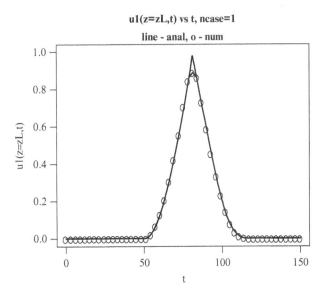

Figure 1.16 Comparison of the numerical and analytical solutions of eqs. (1.1b), (1.2b) ncase=1, ifd=4, pulse BC

$-\omega \sin(\omega(t - z/v))$ and at the peak, $\omega(t - z/v) = \pi/2$ the derivative is $-\omega$. Therefore, the derivative at the peak is a finite step of magnitude 2ω (as demonstrated in Fig. 1.16), and the calculation of $\dfrac{\partial u_1}{\partial z}$ is difficult numerically.

In summary, for ncase=1 the comparison of the numerical and analytical solutions demonstrated smaller differences than for the unit step as expected since the cosine pulse of eq. (1.7a) is smoother than the step of eq. (1.5d). But the differences are large enough that some experimentation for ncase=2 is suggested.

We now compare the numerical solutions for ifd=3,4 and ncase=2 as was done previously for the unit step. Again, this is easily accomplished by replacing the step BC with the pulse BC. The graphical output is in Fig. 1.17 (with points) and Fig. 1.18 (without points).

The differences between 5pbu and van Leer are clear. If they are considered excessive, one possibility to improve the agreement would be to use more points in z since $n = 41$ is a rather coarse grid (h refinement). Another possibility would be to use another flux limiter (e.g., from the set in [1], pp 40-42) which is a generalized form of p refinement in which the algorithm is changed.

The preceding discussion is for a single component with concentrations $u_1(z, t)$ (fluid) and $u_2(z, t)$ (adsorbent). The intention is to demonstrate the features and performance of some approximations for the derivative in $\dfrac{\partial u_1}{\partial z}$ in eq. (1.1b). However, movement of fronts through the chromatographic column (Fig. 1.1) for multi component systems is of primary interest in the application of chromatographic separation and analysis. We therefore next consider how the preceding routines can be extended to two components (and thus, for any number of components).

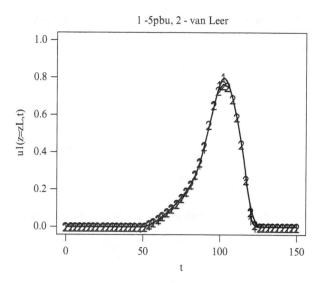

Figure 1.17   Comparison of 5pbu (`ifd=3`) and van Leer (`ifd=4`), `ncase=2`

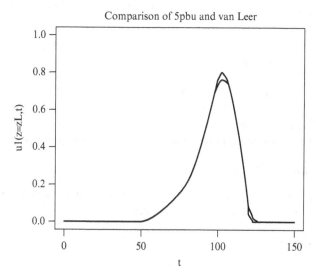

Figure 1.18   Comparison of 5pbu (`ifd=3`) and van Leer (`ifd=4`), `ncase=2`

(1.4) Multi component model

The interest in chromatography is primarily within the context of analysis and separation of multi component mixtures. Here we consider how the previous single component model can be extended to multi component applications.

Eqs. (1.1b), (1.2b) can be extended to a system of two components, with dependent variables u_1, u_2, u_3, u_4 [14]

$$\frac{\partial u_1}{\partial t} = -\frac{\partial(vu_1)}{\partial z} - \frac{(1-\epsilon)}{\epsilon}(k_f u_1(u_2^e - u_2) - k_r u_2) \qquad (1.8a)$$

$$\frac{\partial u_2}{\partial t} = k_f u_1(u_2^e - u_2) - k_r u_2 \qquad (1.8b)$$

$$\frac{\partial u_3}{\partial t} = -\frac{\partial(vu_3)}{\partial z} - \frac{(1-\epsilon)}{\epsilon}(k_f u_3(u_4^e - u_4) - k_r u_4) \qquad (1.8c)$$

$$\frac{\partial u_4}{\partial t} = k_f u_3(u_4^e - u_4) - k_r u_4 \qquad (1.8d)$$

The adsorbent equilibrium concentrations, u_2^e, u_4^e are given by the isotherms

$$u_2^e = \frac{c_1 u_1}{1 + c_2 u_1}; \quad u_4^e = \frac{c_3 u_3}{1 + c_4 u_3} \qquad (1.9a,b)$$

These isotherms are single component, but they could easily be extended to the multi component case, e.g., $u_2^e = f_2(u_1, u_3), u_4^e = f_4(u_1, u_3)$.

The ICs for eqs. (1.8) are

$$u_1(z, t = 0) = f_1(z), \ u_2(z, t = 0) = f_2(z), \ u_3(z, t = 0) = f_3(z),$$
$$u_1(z, t = 0) = f_4(z) \qquad (1.10a,b,c,d)$$

The BCs for eqs. (1.8) are

$$u_1(z = 0, t) = g_1(t), \ u_3(z = 0, t) = g_3(t) \qquad (1.11a,b)$$

We will next consider eqs. (1.8) to (1.11) for homogeneous ICs, $f_1(z) = f_2(z) = f_3(z) = f_4(z) = 0$, and pulse function BCs, $g_1(t) = g_3(t) = \text{pulse}(t)$.

(1.5) MOL routines

The main program and subordinate routines for the multi component model are next. A main program for eqs. (1.8) to (1.11) is in Listing 1.6.

(1.5.1) Main program

```
#
# Delete previous workspaces
  rm(list=ls(all=TRUE))
#
#   1D, two component, chromatography model
#
```

---

[14]We have followed the usual convention of naming PDE dependent variables with u and a number. The alternative is to use variables names that are more closely identified with physical variables, e.g., $c_{f,1}, c_{a,1}, c_{f,2}, c_{a,2}$ where f, a denote fluid and adsorbent, respectively.

```
#    The ODE/PDE system is
#
#    u1_t = -v*u1_z - (1 - eps)/eps*rate1
#
#    u2_t = rate1
#
#    u3_t = -v*u3_z - (1 - eps)/eps*rate3
#
#    u4_t = rate3
#
#    Boundary conditions
#
#       u1(z=0,t) = pulse(t)
#
#       u3(z=0,t) = pulse(t)
#
#    Initial conditions
#
#       u1(z,t=0) = 0
#
#       u2(z,t=0) = 0
#
#       u3(z,t=0) = 0
#
#       u4(z,t=0) = 0
#
#    The method of lines (MOL) solution is coded below.
#    Specifically, the spatial derivatives in the fluid
#    balances, u1_z, u3_z, are replaced by one of four
#    approximations as selected by the variable ifd.
#
# Access ODE integrator
  library("deSolve");
#
# Access files
  setwd("G:/chap1");
  source("pde_1.R") ;source("pulse.R") ;
  source("dss004.R");source("dss012.R");
  source("dss020.R");source("vanl.R")   ;
#
# Step through cases
  for(ncase in 1:2){
#
#    Model parameters
       v=1; eps=0.4;
     u10=0;    u20=0; u30=0; u40=0;
```

```
   c1=1;     c2=1;   c3=1;   c4=1;
  zL=50;    n=41;
  if(ncase==1){ kf1=0; kr2=0; kf3=0; kr4=0;}
  if(ncase==2){ kf1=1; kr2=1; kf3=0; kr4=0;}
#
# Select an approximation for the convective derivative u1z
#
#   ifd = 1: Two point upwind approximation
#
#   ifd = 2: Centered approximation
#
#   ifd = 3: Five point, biased upwind approximation
#
#   ifd = 4: van Leer flux limiter
#
  ifd=3;
#
# Level of output
#
#   Detailed output   - ip = 1
#
#   Brief (IC) output - ip = 2
#
  ip=1;
#
# Initial condition
  u0=rep(0,4*n);
  for(i in 1:n){
    u0[i]=      u10;
    u0[i+n]=   u20;
    u0[i+2*n]=u30;
    u0[i+3*n]=u40;
  }
  t0=0;tf=150;nout=51;
  tout=seq(from=t0,to=tf,by=(tf-t0)/(nout-1));
  ncall=0;
#
# ODE integration
  out=ode(func=pde_1,times=tout,y=u0);
#
# Store solution
  u1=matrix(0,nrow=nout,ncol=n);
  u2=matrix(0,nrow=nout,ncol=n);
  u3=matrix(0,nrow=nout,ncol=n);
  u4=matrix(0,nrow=nout,ncol=n);
  t=rep(0,nout);
```

```
   for(it in 1:nout){
   for(iz in 1:n){
     u1[it,iz]=out[it,iz+1];
     u2[it,iz]=out[it,iz+1+n];
     u3[it,iz]=out[it,iz+1+2*n];
     u4[it,iz]=out[it,iz+1+3*n];
   }
     t[it]=out[it,1];
   }
#
# Display ifd, ncase
   cat(sprintf("\n ifd = %2d   ncase = %2d",ifd,ncase));
#
# Display numerical solution
   if(ip==1){
   cat(sprintf(
     "\n\n      t      u1(z=zL,t)  rate1(z=zL,t)\n"));
   cat(sprintf(
     "\n\n      t      u3(z=zL,t)  rate3(z=zL,t)\n"));
   u2eq=rep(0,nout);  u4eq=rep(0,nout);
   rate1=rep(0,nout);rate3=rep(0,nout);
   for(it in 1:nout){
     u2eq[it]=c1*u1[it,n]/(1+c2*u1[it,n]);
     u4eq[it]=c3*u3[it,n]/(1+c4*u3[it,n]);
     rate1[it]=kf1*u1[it,n]*(u2eq[it]-u2[it,n])-kr2*u2[it,n];
     rate3[it]=kf3*u3[it,n]*(u4eq[it]-u4[it,n])-kr4*u4[it,n];
     cat(sprintf(
       "%7.2f%12.4f%12.4f\n"   ,t[it],u1[it,n],rate1[it]));
     cat(sprintf(
       "%7.2f%12.4f%12.4f\n\n",t[it],u3[it,n],rate3[it]));
   }
   }
#
# Store solution for plotting
   u1plot=rep(0,nout);u3plot=rep(0,nout);
    tplot=rep(0,nout);
   for(it in 1:nout){
     u1plot[it]=u1[it,n];
     u3plot[it]=u3[it,n];
      tplot[it]=t[it];
   }
#
# Calls to ODE routine
   cat(sprintf("\n ncall = %4d\n",ncall));
#
# Plot for u1(z=zL,t)
```

```
# ncase = 1
  if(ncase==1){
  par(mfrow=c(1,1))
  plot(tplot,u1plot,xlab="t",ylab="u1(z=zL,t),u3(z=zL,t)",
    lwd=2,main="u1(z=zL,t),u3(z=zL,t) vs t, ncase=1\n
    line - anal, o - num",type="l",xlim=c(0,tplot[nout]),
    ylim=c(-0.1,1.1));
  points(tplot,u1plot,pch="1",lwd=2);
   lines(tplot,u3plot,lwd=2,type="l");
  points(tplot,u3plot,pch="3",lwd=2);
  }
#
# Analytical solution, ncase=1
  if(ncase==1){
  u1expl=rep(0,nout);u3expl=rep(0,nout);
  for(it in 1:nout){
    u1expl[it]=pulse(tplot[it],zL,v);
    u3expl[it]=pulse(tplot[it],zL,v);
  }
  lines(tplot,u1expl,lwd=2,type="l");
  lines(tplot,u3expl,lwd=2,type="l");
  }
#
# ncase = 2
  if(ncase==2){
  par(mfrow=c(1,1))
  plot(tplot,u1plot,xlab="t",ylab="u1(z=zL,t),u3(z=zL,t)",
    lwd=2,main="u1(z=zL,t),u3(z=zL,t) vs t, ncase=2",
    type="l",xlim=c(0,tplot[nout]),ylim=c(-0.1,1.1));
  points(tplot,u1plot,pch="1",lwd=2);
   lines(tplot,u3plot,lwd=2,type="l");
  points(tplot,u3plot,pch="3",lwd=2);
  }
#
# Next case
  }
```

Listing 1.6: Main program for the multi component model, eqs. (1.8) to (1.11)

Listing 1.6 is similar to Listing 1.1. Therefore, the differences will be emphasized next.

- A block of documentation comments for the two component model, followed by access to the library of ODE solvers deSolve and the routines specific to the coding are at the beginning as before (e.g., in Listings 1.1 and 1.4).
- Two cases are programmed in a for

```
#
# Step through cases
  for(ncase in 1:2){
#
#   Model parameters
      v=1; eps=0.4;
    u10=0;   u20=0; u30=0; u40=0;
     c1=1;      c2=1;   c3=1;   c4=1;
    zL=50;      n=41;
    if(ncase==1){ kf1=0; kr2=0; kf3=0; kr4=0;}
    if(ncase==2){ kf1=1; kr2=1; kf3=0; kr4=0;}
```

For ncase=1, no transfer of the two components to the adsorbent occurs, so the linear advection equation (1.5a) applies to both components. For ncase=2, component 1 is adsorbed while component 2 is not. Thus, we would expect some separation of the two components, that is, differences in $u_1(z = z_L, t)$ and $u_3(z = z_L, t)$. This selective adsorption will be observed in the solutions reported next.

- Four spatial differentiators are again programmed corresponding to ifd=1,2,3,4. The 5pbu FD ifd=3 will be used primarily since it gives little numerical diffusion and oscillation, and is computationally efficient, as observed previously for the one component case.

- Homogeneous (zero) ICs are programmed for $u_1(z, t = 0), u_2(z, t = 0), u_3(z, t = 0), u_4(z, t = 0)$.

```
#
# Initial condition
  u0=rep(0,4*n);
  for(i in 1:n){
    u0[i]=      u10;
    u0[i+n]=    u20;
    u0[i+2*n]=u30;
    u0[i+3*n]=u40;
  }
```

Again n=41 so the number of ODEs in the MOL analysis (of eqs. (1.8a), (1.8b)) is now $4(41) = 164$.

- The variation in t is again $0 \leq t \leq 150$ with 51 output points (including $t = 0$).

```
t0=0;tf=150;nout=51;
tout=seq(from=t0,to=tf,by=(tf-t0)/(nout-1));
ncall=0;
```

- Integrator ode is used to compute the numerical solution in array out. The four dependent variables are then placed in arrays u1,u2,u3,u4 that are 2D for the variations in z and t.

```
#
# ODE integration
  out=ode(func=pde_1,times=tout,y=u0);
#
# Store solution
  u1=matrix(0,nrow=nout,ncol=n);
  u2=matrix(0,nrow=nout,ncol=n);
  u3=matrix(0,nrow=nout,ncol=n);
  u4=matrix(0,nrow=nout,ncol=n);
  t=rep(0,nout);
  for(it in 1:nout){
  for(iz in 1:n){
    u1[it,iz]=out[it,iz+1];
    u2[it,iz]=out[it,iz+1+n];
    u3[it,iz]=out[it,iz+1+2*n];
    u4[it,iz]=out[it,iz+1+3*n];
  }
    t[it]=out[it,1];
  }
```

Also, t (out[it,1]) is placed in vector t as before.

- ifd and ncase are displayed at the beginning of the output. Then, for ip=1 (detailed numerical output), the equilibrium concentrations u_2^e, u_4^e are computed from eqs. (1.9), and the adsorption rates in eqs. (1.8) are computed and placed in vectors and displayed.

```
#
# Display ifd, ncase
  cat(sprintf("\n ifd = %2d   ncase = %2d",ifd,ncase));
#
# Display numerical solution
  if(ip==1){
  cat(sprintf(
    "\n\n      t       u1(z=zL,t)  rate1(z=zL,t)\n"));
  cat(sprintf(
    "\n\n      t       u3(z=zL,t)  rate3(z=zL,t)\n"));
  u2eq=rep(0,nout);  u4eq=rep(0,nout);
  rate1=rep(0,nout);rate3=rep(0,nout);
  for(it in 1:nout){
    u2eq[it]=c1*u1[it,n]/(1+c2*u1[it,n]);
    u4eq[it]=c3*u3[it,n]/(1+c4*u3[it,n]);
    rate1[it]=kf1*u1[it,n]*(u2eq[it]-u2[it,n])-kr2*u2[it,n];
```

```
      rate3[it]=kf3*u3[it,n]*(u4eq[it]-u4[it,n])-kr4*u4[it,n];
      cat(sprintf(
        "%7.2f%12.4f%12.4f\n"   ,t[it],u1[it,n],rate1[it]));
      cat(sprintf(
        "%7.2f%12.4f%12.4f\n\n",t[it],u3[it,n],rate3[it]));
    }
  }
```

- The exiting concentrations $u_1(z = z_L, t), u_3(z = z_L, t)$ are placed in vectors for subsequent plotting (note the use of n corresponding to $z = z_L$).

```
#
# Store solution for plotting
  u1plot=rep(0,nout);u3plot=rep(0,nout);
   tplot=rep(0,nout);
  for(it in 1:nout){
    u1plot[it]=u1[it,n];
    u3plot[it]=u3[it,n];
     tplot[it]=t[it];
  }
```

- The number of calls to the MOL/ODE routine pde_1 (discussed next) is displayed. Then $u_1(z = z_L, t), u_3(z = z_L, t)$ are plotted against t and identified with the characters 1,3 in the plot using the utilities plot, points, lines.

```
#
# Calls to ODE routine
  cat(sprintf("\n ncall = %4d\n",ncall));
#
# Plot for u1(z=zL,t)
# ncase = 1
  if(ncase==1){
  par(mfrow=c(1,1))
  plot(tplot,u1plot,xlab="t",ylab="u1(z=zL,t),u3(z=zL,t)",
    lwd=2,main="u1(z=zL,t),u3(z=zL,t) vs t, ncase=1\n
    line - anal, o - num",type="l",xlim=c(0,tplot[nout]),
    ylim=c(-0.1,1.1));
  points(tplot,u1plot,pch="1",lwd=2);
   lines(tplot,u3plot,lwd=2,type="l");
  points(tplot,u3plot,pch="3",lwd=2);
  }
```

- For ncase=1 (no adsorption), the analytical solution of eqs. (1.8) is placed in two arrays by a call to pulse of Listing 1.5.

```
#
# Analytical solution, ncase=1
  if(ncase==1){
  u1expl=rep(0,nout);u3expl=rep(0,nout);
  for(it in 1:nout){
    u1expl[it]=pulse(tplot[it],zL,v);
    u3expl[it]=pulse(tplot[it],zL,v);
  }
  lines(tplot,u1expl,lwd=2,type="l");
  lines(tplot,u3expl,lwd=2,type="l");
  }
```

The analytical solutions are then superimposed on the ncase=1 plot with the lines utility. The superposition takes place because the par(mfrow=c(1,1)) is not repeated (for a separate plot).

- For ncase=2 (with adsorption of component 1), the solutions $u_1(z = z_L, t), u_3(z = z_L, t)$ are plotted as lines with points.

```
#
# ncase = 2
  if(ncase==2){
  par(mfrow=c(1,1))
  plot(tplot,u1plot,xlab="t",ylab="u1(z=zL,t),u3(z=zL,t)",
    lwd=2,main="u1(z=zL,t),u3(z=zL,t) vs t, ncase=2",
    type="l",xlim=c(0,tplot[nout]),ylim=c(-0.1,1.1));
  points(tplot,u1plot,pch="1",lwd=2);
   lines(tplot,u3plot,lwd=2,type="l");
  points(tplot,u3plot,pch="3",lwd=2);
  }
#
# Next case
  }
```

The final } concludes the for in ncase.

(1.5.2) MOL/ODE routine

The MOL/PDE routine pde_1 called by ode in Listing 1.7 is next.

```
  pde_1=function(t,u,parms) {
#
# Function pde_1 computes the t derivative vector of the u vector
#
# One vector to four PDEs
  u1=rep(0,n);u2=rep(0,n);
  u3=rep(0,n);u4=rep(0,n);
```

```
  for (i in 1:n){
    u1[i]=u[i];
    u2[i]=u[i+n];
    u3[i]=u[i+2*n];
    u4[i]=u[i+3*n];
  }
#
# Boundary condition
  u1[1]=pulse(t,0,v);
  u3[1]=pulse(t,0,v);
#
# First order spatial derivative
#
#   ifd = 1: Two point upwind finite difference (2pu)
    if(ifd==1){ u1z=dss012(0,zL,n,u1,v); }
    if(ifd==1){ u3z=dss012(0,zL,n,u3,v); }
#
#   ifd = 2: Five point center finite difference (5pc)
    if(ifd==2){ u1z=dss004(0,zL,n,u1); }
    if(ifd==2){ u3z=dss004(0,zL,n,u3); }
#
#   ifd = 3: Five point biased upwind approximation (5pbu)
    if(ifd==3){ u1z=dss020(0,zL,n,u1,v); }
    if(ifd==3){ u3z=dss020(0,zL,n,u3,v); }
#
#   ifd = 4: van Leer flux limiter
    if(ifd==4){ u1z=vanl(0,zL,n,u1,v); }
    if(ifd==4){ u3z=vanl(0,zL,n,u3,v); }
#
# Temporal derivatives, mass transfer rate
    u1t=rep(0,n);   u2t=rep(0,n);
    u3t=rep(0,n);   u4t=rep(0,n);
  u2eq=rep(0,n);rate1=rep(0,n);
  u4eq=rep(0,n);rate3=rep(0,n);
#
#   u1t, u2t
    for(i in 1:n){
      u2eq[i]=c1*u1[i]/(1+c2*u1[i]);
      rate1[i]=kf1*u1[i]*(u2eq[i]-u2[i])-kr2*u2[i];
      if(i==1){
        u1t[i]=0;
      }else{
        u1t[i]=-v*u1z[i]-(1-eps)/eps*rate1[i];
      }
        u2t[i]=rate1[i];
    }
```

```
#
#    u3t, u4t
     for(i in 1:n){
        u4eq[i]=c3*u3[i]/(1+c4*u3[i]);
        rate3[i]=kf3*u3[i]*(u4eq[i]-u4[i])-kr4*u4[i];
        if(i==1){
           u3t[i]=0;
        }else{
           u3t[i]=-v*u3z[i]-(1-eps)/eps*rate3[i];
        }
           u4t[i]=rate3[i];
     }
#
# Four PDEs to one vector
  ut=rep(0,4*n);
  for(i in 1:n){
    ut[i]     =u1t[i];
    ut[i+n]   =u2t[i];
    ut[i+2*n]=u3t[i];
    ut[i+3*n]=u4t[i];
  }
#
# Increment calls to pde_1
  ncall<<-ncall+1;
#
# Return derivative vector
  return(list(c(ut)));
}
```

Listing 1.7: MOL/ODE routine pde_1 for eqs. (1.8), (1.9) and (1.11)

Listing 1.7 is similar to Listing 1.2 so only the differences are emphasized next.

- The function is first defined. Then the input vector u of length $4(41) = 164$ is placed in four vectors of length 41.

```
    pde_1=function(t,u,parms) {
#
# Function pde_1 computes the t derivative vector
# of the u vector
#
# One vector to four PDEs
    u1=rep(0,n);u2=rep(0,n);
    u3=rep(0,n);u4=rep(0,n);
    for (i in 1:n){
       u1[i]=u[i];
```

```
        u2[i]=u[i+n];
        u3[i]=u[i+2*n];
        u4[i]=u[i+3*n];
    }
```

This use of four arrays facilitates the programming of eqs. (1.8) (four PDEs).

- A pulse BC is used at $z = 0$ for both eqs. (1.8a) and (1.8c) for ncase=1,2.

```
#
# Boundary condition
    u1[1]=pulse(t,0,v);
    u3[1]=pulse(t,0,v);
```

In other words, this programming is for BCs (1.11) with a pulse. Since the programming is the same for eqs. (1.8a) and (1.8c), the numerical solutions should be the same (for ncase=1), which serves as a check on the coding in pde_1 of Listing (1.6).

- The derivatives in z in eqs. (1.8a) and (1.8c) are computed by one of the four spatial differentiators considered previously (in Listing 1.2). For the model of eqs. (1.8), ifd=3 is used, for which the programming is

```
#
#    ifd = 3: Five point biased upwind approximation (5pbu)
    if(ifd==3){ u1z=dss020(0,zL,n,u1,v);  }
    if(ifd==3){ u3z=dss020(0,zL,n,u3,v);  }
```

The two derivatives $\dfrac{\partial u_1}{\partial z}, \dfrac{\partial u_3}{\partial z}$ in eqs. (1.8a) and (1.8c) are calculated in this way.

- Arrays are declared for the derivatives in t in eqs. (1.8a), (1.8c), the equilibrium concentrations of eqs. (1.9), and the adsorption rates in eqs. (1.8a) and (1.8c).

```
#
# Temporal derivatives, equilibrium concentrations,
# mass transfer rates
    u1t=rep(0,n);   u2t=rep(0,n);
    u3t=rep(0,n);   u4t=rep(0,n);
  u2eq=rep(0,n);rate1=rep(0,n);
  u4eq=rep(0,n);rate3=rep(0,n);
```

- Eq. (1.8a) is programmed as

```
#
#    u1t, u2t
    for(i in 1:n){
      u2eq[i]=c1*u1[i]/(1+c2*u1[i]);
      rate1[i]=kf1*u1[i]*(u2eq[i]-u2[i])-kr2*u2[i];
      if(i==1){
```

```
          u1t[i]=0;
        }else{
          u1t[i]=-v*u1z[i]-(1-eps)/eps*rate1[i];
        }
          u2t[i]=rate1[i];
      }
```

A `for` is used for the interval $0 \leq z \leq z_L$. The equilibrium concentration u_2^e, and adsorption rate are calculated and placed in vectors. For $z = 0$, since the entering concentration is defined by BC (1.11a), the derivative in t is set to zero to retain this boundary value. For $z > 0$, eqs. (1.8a), (1.8b) are used to compute the derivatives $\dfrac{\partial u_1}{\partial t}, \dfrac{\partial u_2}{\partial t}$.

- In the same way, the derivatives $\dfrac{\partial u_3}{\partial t}, \dfrac{\partial u_4}{\partial t}$ in eqs. (1.8c), (1.8d) are computed.

```
#
#     u3t, u4t
      for(i in 1:n){
        u4eq[i]=c3*u3[i]/(1+c4*u3[i]);
        rate3[i]=kf3*u3[i]*(u4eq[i]-u4[i])-kr4*u4[i];
        if(i==1){
          u3t[i]=0;
        }else{
          u3t[i]=-v*u3z[i]-(1-eps)/eps*rate3[i];
        }
          u4t[i]=rate3[i];
      }
```

- The four derivatives in t are placed in a single vector, ut, to return to the ODE integrator, ode (called in Listing 1.6).

```
#
# Four PDEs to one vector
  ut=rep(0,4*n);
  for(i in 1:n){
    ut[i]    =u1t[i];
    ut[i+n]  =u2t[i];
    ut[i+2*n]=u3t[i];
    ut[i+3*n]=u4t[i];
  }
```

- The counter for the calls to pde_1 is incremented and returned to the main program of Listing 1.6 with <<-.

```
#
# Increment calls to pde_1
  ncall<<-ncall+1;
```

- The vector ut is returned to ode with a combination of c (the R vector operator), list (required by ode), and return.

```
#
# Return derivative vector
  return(list(c(ut)));
}
```

The final } concludes the function pde_1.

The only other required function is for the BCs of eqs. (1.11), pulse in this case in Listing 1.5. The output from the routines in Listings 1.6, 1.7 and 1.5 is considered next.

(1.6) Model output, multi component chromatography

The numerical output from Listing 1.6 is not reproduced here to conserve space. The two plots produced by the main program of Listing 1.6 are in Figs. 1.19 and 1.20.

In Fig. 1.19, all four solutions are essentially identical (for 5pbu, $u_1(t, z = z_L)$, $u_3(t, z = z_L)$, numerical and analytical) as expected. This is a worthwhile check on the coding since any errors might produce different solutions.

In Fig. 1.20, the numerical solutions reflect the difference of adsorption for the two components, i.e., component 1 is adsorbed (from the values kf1=1, kr1=1 in Listing 1.6), while component 2 is not adsorbed (from the values kf2=0, kr2=0). As expected, component 2 leaves the column first and a partial separation is effected. While the output

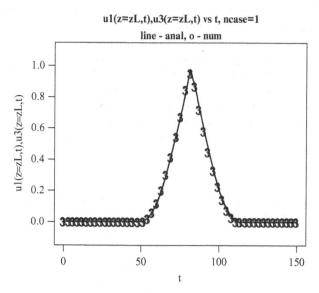

Figure 1.19 Comparison of the numerical and analytical solutions of eqs. (1.8) ncase=1, ifd=3, pulse BC

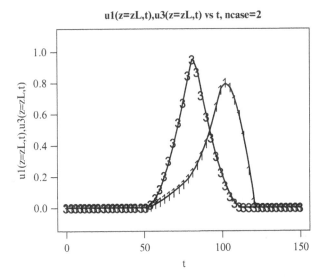

Figure 1.20   Numerical solutions of eqs. (1.8), `ncase=2`, `ifd=3`, pulse BC

stream is not very pure in either component, the purity of the output stream could be substantially enhanced by, for example, increasing the length of the column, changing the velocity v, using multiple columns in sequence, using an adsorbent with different selective properties, etc.

These various options can be studied with the model which would be time consuming and expensive experimentally. Also, additional components can easily be added to the coding in Listings 1.6 and 1.7. This type of design study and possible optimization illustrates the inherent value of a mathematical model.

We now consider a series of PDE models for BMSE applications in subsequent chapters. When considering the models, we will introduce diffusion modeled by second derivatives in the spatial derivatives (*parabolic* PDEs), that is, the use of Fick's first and second laws, including nonlinear extensions. However, the intent in this chapter is to focus on the inherent difficulties of solving PDE models numerically for strongly convective (hyperbolic) systems. The addition of diffusion eases the numerical requirements since steep fronts and discontinuities are smoothed by diffusion.

Reference

[1] Griffiths, G.W., and W.E. Schiesser (2012), *Traveling Wave Analysis of Partial Differential Equations*, Elsevier/Academic Press, Boston, MA

2

WAVE FRONT RESOLUTION IN VEGF ANGIOGENESIS

The central topic of this chapter is the generation of new blood vessels, generally termed *angiogenesis*, in response to a growth factor, e.g., vascular endothelium growth factor or VEGF. The dynamics of VEGF production, and its effect on endothelium cell growth, is modeled by two simultaneous (coupled) nonlinear PDEs. Explanatory discussion of these PDEs, followed by a MOL numerical solution and a special-case analytical solution, are presented in the format of Chapter 1.

The following excerpt from [5] provides some additional background:

> Angiogenesis is a chemostatic[1] process involving the generation of new blood vessels from pre-existing vessels. It is essential for the growth and development of solid tumors as well as cancer metastasis. Tumor angiogenesis starts with cancerous tumor cells secreting some chemical substance (or signalling molecule), which is generally called angiogenesis growth factor, to induce neighboring endothelial cells to migrate toward the tumor in order to build its own capillary network to supply nutrients and oxygen for its development.

Introductory background information pertaining to VEGF angiogenesis is also available from the National Cancer Institute, including instructive figures [3].

The intent of this chapter is to

- Discuss a 1D 2-PDE model, including the required initial conditions (ICs) and boundary conditions (BCs). The model has some particular features that will be emphasized.

---

[1]Chemotaxis is explained and modeled in [4], Chapter 1.

Method of Lines PDE Analysis in Biomedical Science and Engineering, First Edition. William E. Schiesser.
© 2016 John Wiley & Sons, Inc. Published 2016 by John Wiley & Sons, Inc.
Companion website: www.wiley.com/go/Schiesser/PDE_Analysis

- – Second-order spatial derivatives to model diffusion. These second-order derivatives are computed by successive first-order numerical differentiation.
- – A traveling-wave solution altered by the diffusion.
- – Nonlinear diffusion and reaction terms to model angiogenesis.
- • Illustrate the coding of the model through a series of R routines, including the use of library routines for integration of the PDE derivatives in time and space.
- • Use a special case analytical solution to evaluate the numerical MOL solution.
- • Present the computed model solutions in numerical and graphical (plotted) formats.
- • Discuss the features of the numerical solution.

(2.1) 1D 2-PDE model

The PDE model for VEGF angiogenesis is[2]

$$u_{1t} = \epsilon u_{1xx} - k u_1^{m_1} u_2^{m_2} \tag{2.1a}$$

$$u_{2t} = d(u_{2x} - \chi u_1^{-\alpha} u_2 u_{1x})_x \tag{2.1b}$$

$$u_1(x, t = 0) = f_1(x); \quad u_2(x, t = 0) = f_2(x) \tag{2.2a,b}$$

$$u_{1x}(x \to -\infty, t) = u_{1x}(x \to \infty, t) = 0 \tag{2.3a,b}$$

$$u_{2x}(x \to -\infty, t) = u_{2x}(x \to \infty, t) = 0 \tag{2.3c,d}$$

where

$u_1(x, t)$	concentration of angiogenesis growth factor (VEGF)
$u_2(x, t)$	density of endothelial cells
x	distance (boundary value independent variable)
t	time (initial value independent variable)
$f_1(x), f_2(x)$	initial condition functions
$d, k, m_1, m_2,$	specified constants
χ, α, ϵ	

Eqs. (2.3) are termed *Neumann* BCs since they specify the first-order derivative of the PDE dependent variables (u_1, u_2) with respect to the spatial independent variable x.

---

[2]Subscript notation is used for the partial derivatives. For example,

$$\frac{\partial u_1}{\partial t} \Rightarrow u_{1t}; \quad \frac{\partial u_1}{\partial x} \Rightarrow u_{1x}; \quad \frac{\partial u_2}{\partial t} \Rightarrow u_{2t}; \quad \frac{\partial^2 u_1}{\partial x^2} \Rightarrow u_{1xx}$$

The subscript notation is used because of its simplicity (relative to the usual partial derivatives with ∂), and its close resemblance to the corresponding programming, e.g., u_{1t}, is programmed as u1t.

Specific values of $d, k, m_1, m_2, \chi, \alpha, \epsilon$ are indicated in the subsequent discussion of the R routines. For the special case, $k = 1, m_1 = 0, m_2 = 1, \chi = 2, \alpha = 1, \epsilon = 0$, the analytical solution to eqs. (2.1), (2.2), and (2.3) is ([2], p68)

$$u_1^a(z) = \left[1 + e^{-cz/d}\right]^{-1} \tag{2.4a}$$

$$u_2^a(z) = \frac{c^2}{kd} e^{-cz/d} \left[1 + e^{-cz/d}\right]^{-2} \tag{2.4b}$$

where $z = x - ct$ and c is a constant to be specified (a velocity). Note that $u_1(z), u_2(z)$ are a function of the single Lagrangian variable z. In other words, these solutions are invariant for a constant value of z, regardless of how x and t may vary. A solution with this property is termed a traveling wave as discussed in Chapter 1.

Eqs. (2.4) are used subsequently to evaluate the numerical solutions of eqs. (2.1) to (2.3) with the ICs of eqs. (2.2), that is, for $t = 0$ and therefore $z = x - ct = x$

$$u_1(x, t = 0) = u_1^a(x); \quad u_2(x, t = 0) = u_2^a(x) \tag{2.5a,b}$$

We can note the following details about eqs. (2.1) to (2.5).

- Second-order derivatives in x model diffusion[3]. These second-order derivatives originate with *Fick's first and second laws*.
- Since eqs. (2.1) are second order in x, they each require two BCs. Eqs. (2.3) are the pairs of BCs for eqs. (2.1). Also, since eqs. (2.1) are first order in t, each requires one IC, that is, eqs. (2.2).
- The BCs can be defined on an infinite interval, e.g., $-\infty \le x \le \infty$ as for eqs. (2.3); a semi-infinite interval, e.g., $0 \le x \le \infty$; or a finite interval, e.g., $0 \le x \le x_L$ (as in Chapter 1). Since $\pm\infty$ cannot be represented on a computer (with finite arithmetic), finite boundary values are used that are effectively infinite. This procedure for approximating BCs (2.3) is illustrated in the computer analysis that follows.
- Eqs. (2.1) are coupled through terms that include both u_1 and u_2, such as $-k u_1^{m_1} u_2^{m_2}$ and $(\chi u_1^{-\alpha} u_2 u_{1x})_x$. That is, eqs. (2.1) must be solved together or simultaneously.
- Eqs. (2.1) have several forms of nonlinearity[4]. The term $-k u_1^{m_1} u_2^{m_2}$ in eq. (2.1a) has a product of the two dependent variables u_1 and u_2, raised to powers m_1, m_2 (the term is nonlinear, even for $m_1 = m_2 = 1$). The term $(\chi u_1^{-\alpha} u_2 u_{1x})_x$ in

---

[3]PDEs that include: (1) first-order derivatives in x for convection as in Chapter 1, (2) second-order derivatives in x for diffusion as in eqs. (2.1), and (3) reaction terms as in eq. (2.1a) $(-k u_1^{m_1} u_2^{m_2})$ are termed *convection-diffusion-reaction* (CDR) PDEs. 3D CDR PDEs are derived in Appendix A.

[4]A PDE is linear if its dependent variable, e.g., u_1 of eq. (2.1a), and all of its derivatives are to the first power or first *degree*. As a word of caution, degree should not be confused with order. The order is determined by the highest order derivative in the PDE. For example, eq. (2.1a) is first order in t, determined by the highest order derivative, u_{1t}, in t. It is second order in x, as determined by the highest order derivative, u_{1xx}, in x. It is not first degree (or linear) because of the nonlinear term $(\chi u_1^{-\alpha} u_2 u_{1x})_x$ and, in particular, the product $u_1^{-\alpha} u_2 u_{1x}$ (which is nonlinear even for $\alpha = 0$). Generally, nonlinear PDEs are more difficult to integrate analytically than linear PDEs. The availability of the analytical solution eqs. (2.4) is unusual.

eq. (2.1b) has the product $u_1^{-\alpha}u_2u_{1x}$ that involves the dependent variables u_1, u_2 and a derivative u_{1x}.

- The nonlinear term in eq. (2.1b), $(\chi u_1^{-\alpha}u_2u_{1x})_x$, is second order (note the outer differentiation of the inner first derivative) and represents a form of diffusion to model chemotaxis [4]. The implementation of this term is illustrated in the coding that follows.

Eqs. (2.1) to (2.5) are the complete statement of the model, including the analytical solution of eqs. (2.4) that will be used to evaluate the numerical MOL solution. The programming of these equations is considered next.

(2.2) MOL routines

The MOL routines follow the format of the analogous routines in Chapter 1.

(2.2.1) Main program

The main program is in Listing 2.1.

```
#
# Delete previous workspaces
  rm(list=ls(all=TRUE))
#
# Two PDE angiogenesis model
#
# Access ODE integrator
  library("deSolve");
#
# Access functions for numerical, analytical solutions
  setwd("g:/chap2");
  source("angio_1.R");source("dss004.R" );
  source("u1_anal.R");source("u2_anal.R");
#
# Level of output
#
#   ip = 1 - graphical (plotted) solutions
#              (u1(x,t), u2(x,t)) only
#
#   ip = 2 - numerical and graphical solutions
#
  ip=1;
#
# Step through ncase
  for(ncase in 1:2){
#
```

```
# Grid (in x)
  n=101;xl=-10;xu=15
  x=seq(from=xl,to=xu,by=(xu-xl)/(n-1));
#
# Parameters
  c=1;d=1;k=1;m1=0;m2=1;chi=2;alpha=1;
  if(ncase==1){eps=0;}
  if(ncase==2){eps=1;}
  cat(sprintf("\n ncase = %5d\n",ncase));
  cat(sprintf("\n     c = %5.2f    d = %5.2f    k = %5.2f\n",
              c,d,k));
  cat(sprintf("\n    m1 = %5.2f   m2 = %5.2f chi = %5.2f\n",
              m1,m2,chi));
  cat(sprintf("\n alpha = %5.2f eps = %5.2f\n",alpha,eps));
#
# Independent variable for ODE integration
  nout=6;t0=0;tf=5;
  tout=seq(from=0,to=tf,by=(tf-t0)/(nout-1));
#
# Initial condition (from analytical solutions,t=0)
  u0=rep(0,2*n);
  for(i in 1:n){
    u0[i]  =u1_anal(x[i],tout[1],k,d,c);
    u0[i+n]=u2_anal(x[i],tout[1],k,d,c);
  }
  ncall=0;
#
# ODE integration
  out=lsodes(func=angio_1,y=u0,times=tout,parms=NULL)
  nrow(out)
  ncol(out)
#
# Arrays for plotting numerical, analytical solutions
  u1_plot=matrix(0,nrow=n,ncol=nout);
  u2_plot=matrix(0,nrow=n,ncol=nout);
 u1a_plot=matrix(0,nrow=n,ncol=nout);
 u2a_plot=matrix(0,nrow=n,ncol=nout);
  for(it in 1:nout){
    for(ix in 1:n){
       u1_plot[ix,it]=out[it,ix+1];
       u2_plot[ix,it]=out[it,ix+1+n];
      if(ncase==1){
       u1a_plot[ix,it]=u1_anal(x[ix],tout[it],k,d,c);
       u2a_plot[ix,it]=u2_anal(x[ix],tout[it],k,d,c);
      }
    }
```

```
  }
#
# Display numerical, analytical solutions
  if((ncase==1)&(ip==2)){
    for(it in 1:nout){
      cat(sprintf("\n     t        x   u1(x,t)  u1_ex(x,t)
          u1_err(x,t)"));
      cat(sprintf("\n                    u2(x,t)  u2_ex(x,t)
          u2_err(x,t)\n"));
      for(ix in 1:n){
        cat(sprintf("%5.1f%8.2f%10.5f%12.5f%13.6f\n",
                    tout[it],x[ix],
                    u1_plot[ix,it],u1a_plot[ix,it],
                    u1_plot[ix,it]-u1a_plot[ix,it]));
        cat(sprintf("              %10.5f%12.5f%13.6f\n",
        u2_plot[ix,it],u2a_plot[ix,it],
                    u2_plot[ix,it]-u2a_plot[ix,it]));
      }
    }
  }
  if((ncase==2)&(ip==2)){
    for(it in 1:nout){
      cat(sprintf("\n     t        x   u1(x,t)   u2(x,t)"));
      for(ix in 1:n){
        cat(sprintf("%5.1f%8.2f%10.5f%10.5f\n",tout[it],x[ix],
        u1_plot[ix,it],u2_plot[ix,it]));
      }
    }
  }
#
# Calls to ODE routine
  cat(sprintf("\n\n ncall = %5d\n\n",ncall));
#
# Plot u1 numerical, analytical solutions
  if(ncase==1){
  par(mfrow=c(1,1));
  matplot(x=x,y=u1_plot,type="l",xlab="x",ylab="u1(x,t),
          t=0,1,2,3,4,5",xlim=c(xl,xu),lty=1,main="u1(x,t);
          solid - num, points - anal;t=0,1,2,3,4,5;",
          col="black",lwd=2);
  matpoints(x=x,y=u1a_plot,xlim=c(xl,xu),col="black",lwd=2)
#
# Plot u2 numerical, analytical
  par(mfrow=c(1,1));
  matplot(x=x,y=u2_plot,type="l",xlab="x",ylab="u2(x,t),
          t=0,1,2,3,4,5",xlim=c(xl,xu),lty=1,main="u2(x,t),
```

```
                solid - num, points - anal;t=0,1,2,3,4,5;",
                col="black",lwd=2);
      matpoints(x=x,y=u2a_plot,xlim=c(xl,xu),col="black",lwd=2)
      }
      if(ncase==2){
      par(mfrow=c(1,1));
      matplot(x=x,y=u1_plot,type="l",xlab="x",ylab="u1(x,t),
                t=0,1,2,3,4,5",xlim=c(xl,xu),lty=1,main="u1(x,t);
                t=0,1,2,3,4,5;",col="black",lwd=2);
      par(mfrow=c(1,1));
      matplot(x=x,y=u2_plot,type="l",xlab="x",ylab="u2(x,t),
                t=0,1,2,3,4,5",xlim=c(xl,xu),lty=1,main="u2(x,t);
                t=0,1,2,3,4,5;",col="black",lwd=2);
      }
#
# Next case
      }
```

Listing 2.1: Main program for Eqs. (2.1) to (2.5)

We can note the following details about Listing 2.1.

- Previous files are cleared.

```
  #
  # Delete previous workspaces
    rm(list=ls(all=TRUE))
```

- The ODE integrator library deSolve and the files for the model of eqs. (2.1) to (2.5) are accessed.

```
  #
  # Access functions for numerical, analytical solutions
    setwd("g:/chap2");
    source("angio_1.R");source("dss004.R" );
    source("u1_anal.R");source("u2_anal.R");
```

The setwd (set working directory) requires editing for the local computer (note the use of / rather than the usual \).

- The level of output is selected.

```
  #
  # Level of output
  #
  #    ip = 1 - graphical (plotted) solutions
  #               (u1(x,t), u2(x,t)) only
```

```
#
#   ip = 2 - numerical and graphical solutions
#
   ip=1;
```

With ip=1, graphical output only is selected to conserve space. But, a comparison of the analytical and numerical solutions with ip=2 is instructive.

- Two cases are programmed that are discussed later.

```
#
# Step through ncase
   for(ncase in 1:2){
#
```

- A MOL grid of 101 points in x is defined for $-10 \le x \le 15$. This grid, with spacing $(xu-xl)/(n-1)=(15-(-10))/(101-1)=0.25$, was selected to accommodate BCs (2.3). That is, $x = -10, -9.75,...,15$ is effectively an infinite interval $-\infty \le x \le \infty$ as will be observed in the computed solutions.

```
#
# Grid (in x)
   n=101;xl=-10;xu=15
   x=seq(from=xl,to=xu,by=(xu-xl)/(n-1));
```

The seq utility produces the sequence in x.

- The parameters of eqs. (2.1) are defined numerically. For ncase=1, the diffusivity eps in eq. (2.1a) is zero, while for ncase=2, it is one. The change in this parameter (thereby introducing diffusion into eq. (2.1a)) completely changes the form of the model solution as explained in the subsequent discussion.

```
#
# Parameters
   c=1;d=1;k=1;m1=0;m2=1;chi=2;alpha=1;
   if(ncase==1){eps=0;}
   if(ncase==2){eps=1;}
   cat(sprintf("\n ncase = %5d\n",ncase));
   cat(sprintf("\n     c = %5.2f   d = %5.2f   k = %5.2f\n",
               c,d,k));
   cat(sprintf("\n    m1 = %5.2f  m2 = %5.2f chi = %5.2f\n",
               m1,m2,chi));
   cat(sprintf("\n alpha = %5.2f eps = %5.2f\n",alpha,eps));
```

- The interval in t is $0 \le t \le 5$ with 6 output points (including $t = 0$) so that $t = 0, 1,...,5$.

```
#
# Independent variable for ODE integration
   nout=6;t0=0;tf=5;
   tout=seq(from=t0,to=tf,by=(tf-t0)/(nout-1));
```

- For ICs (2.5), functions u1_anal,u2_anal are used, which implement the analytical solutions of eqs. (2.4). tout[1] is the initial value of t, that is, $t = 0$. x[i] varies x in a for according to ICs (2.5). k,d,c are parameters (constants) defined previously.

```
#
# Initial condition (from analytical solutions, t=0)
  u0=rep(0,2*n);
  for(i in 1:n){
    u0[i]  =u1_anal(x[i],tout[1],k,d,c);
    u0[i+n]=u2_anal(x[i],tout[1],k,d,c);
  }
  ncall=0;
```

The initial condition vector u0 has $2(41) = 82$ elements. Functions u1_anal,u2_anal are discussed subsequently. The number of calls to the MOL/ODE routine angio_1 (discussed next) is initialized.

- The ODE system is integrated with lsodes,[5] one of the integrators in deSolve. As expected, the arguments to lsodes are (1) angio_1, the MOL/ODE routine, (2) the IC vector u0 (that informs lsodes of the number of ODEs via its length (82)), and tout, the vector of output values of t. parms for passing parameters to angio_1 is unused.

  The solution matrix out returned by lsodes is dimensioned as out[nout,2*n+1] =out[6,83]. The second dimension is 2*n+1 so that the output values of t are also included (as out[nout,1], the same values of t as in tout).

  The dimensions of out are displayed with the utilities nrow,ncol just to confirm the expected values (6,83).

```
#
# ODE integration
  out=lsodes(func=angio_1,y=u0,times=tout,parms=NULL)
  nrow(out)
  ncol(out)
```

- The numerical solution in out is placed in two matrices for subsequent plotting. Two nested for's are used to step through t and x. For ncase=1, the analytical solution of eqs. (2.4) is also placed in matrices for comparison with the numerical solution and subsequent plotting.

```
#
# Arrays for plotting numerical, analytical solutions
  u1_plot=matrix(0,nrow=n,ncol=nout);
  u2_plot=matrix(0,nrow=n,ncol=nout);
 u1a_plot=matrix(0,nrow=n,ncol=nout);
```

---

[5]lsodes ⇒ Livermore Solver for Ordinary Differential Equations, Sparse, is a quality ODE integrator that is based on sparse matrix methods to conserve storage. It has many options, e.g., error tolerances, minimum and maximum integration steps, and here, only the defaults are used.

```
u2a_plot=matrix(0,nrow=n,ncol=nout);
  for(it in 1:nout){
    for(ix in 1:n){
      u1_plot[ix,it]=out[it,ix+1];
      u2_plot[ix,it]=out[it,ix+1+n];
      if(ncase==1){
        u1a_plot[ix,it]=u1_anal(x[ix],tout[it],k,d,c);
        u2a_plot[ix,it]=u2_anal(x[ix],tout[it],k,d,c);
      }
    }
  }
```

Note the use of functions u1_anal, u2_anal for $-10 \leq x \leq 15, 0 \leq t \leq 5$.

- For ncase=1 and ip=2, the numerical solution of eqs. (2.1), $u_1(x,t)$ and $u_2(x,t)$, and the analytical solution of eqs. (2.4), $u_1^a(x,t)$ and $u_2^a(x,t)$, are displayed as a function of t and x through two nested for's.

```
#
# Display numerical, analytical solutions
  if((ncase==1)&(ip==2)){
    for(it in 1:nout){
      cat(sprintf("\n    t        x    u1(x,t)   u1_ex(x,t)
          u1_err(x,t)"));
      cat(sprintf("\n                     u2(x,t)   u2_ex(x,t)
          u2_err(x,t)\n"));
      for(ix in 1:n){
        cat(sprintf("%5.1f%8.2f%10.5f%12.5f%13.6f\n",
                      tout[it],x[ix],
                      u1_plot[ix,it],u1a_plot[ix,it],
                      u1_plot[ix,it]-u1a_plot[ix,it]));
        cat(sprintf("              %10.5f%12.5f%13.6f\n",
          u2_plot[ix,it],u2a_plot[ix,it],
                      u2_plot[ix,it]-u2a_plot[ix,it]));
      }
    }
  }
```

Note the use of the analytical solution (eqs. (2.4)) for the calculation of the error in the numerical solution.

- For ncase=2 and ip=2, the numerical solution of eqs. (2.1), $u_1(x,t)$ and $u_2(x,t)$, is displayed as a function of t and x through two nested for's (an analytical solution is not available for this case).

```
if((ncase==2)&(ip==2)){
  for(it in 1:nout){
    cat(sprintf("\n    t        x    u1(x,t)   u2(x,t)"));
```

```
      for(ix in 1:n){
        cat(sprintf("%5.1f%8.2f%10.5f%10.5f\n",tout[it],x[ix],
        u1_plot[ix,it],u2_plot[ix,it]));
      }
    }
  }
```

- The number of calls to angio_1 is displayed as a measure of the computational effort required to calculate the numerical solution.

```
#
# Calls to ODE routine
  cat(sprintf("\n\n ncall = %5d\n\n",ncall));
```

- For ncase=1, the numerical solution is plotted as solid lines and the analytical solution is plotted as points (superimposed), using the utilities matplot, matpoints. The arguments of these utilities are straightforward and do not require explanation. They are also illustrated in Figs. 2.1 and 2.2 (discussed subsequently).

```
#
# Plot u1 numerical, analytical solutions
  if(ncase==1){
  par(mfrow=c(1,1));
  matplot(x=x,y=u1_plot,type="l",xlab="x",ylab="u1(x,t),
          t=0,1,2,3,4,5",xlim=c(xl,xu),lty=1,main="u1(x,t);
          solid - num, points - anal;t=0,1,2,3,4,5;",
          col="black",lwd=2);
```

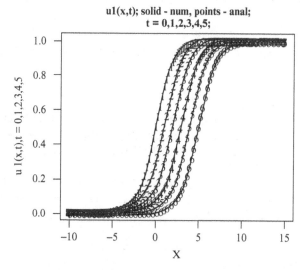

Figure 2.1 Numerical and analytical solutions of eq. (2.1a), $u_1(x,t)$, $u_1^a(x,t)$, (ncase=1)

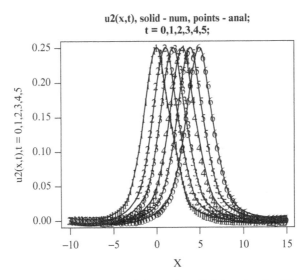

Figure 2.2 Numerical and analytical solutions of eq. (2.1a), $u_2(x,t)$, $u_2^a(x,t)$, (ncase=1)

```
matpoints(x=x,y=u1a_plot,xlim=c(xl,xu),col="black",lwd=2)
#
# Plot u2 numerical, analytical
par(mfrow=c(1,1));
matplot(x=x,y=u2_plot,type="l",xlab="x",ylab="u2(x,t),
        t=0,1,2,3,4,5",xlim=c(xl,xu),lty=1,main="u2(x,t),
        solid - num, points - anal;t=0,1,2,3,4,5;",
        col="black",lwd=2);
matpoints(x=x,y=u2a_plot,xlim=c(xl,xu),col="black",lwd=2)
}
```

- For ncase=2, only the numerical solution is plotted (to produce Figs. 2.3 and 2.4).

```
if(ncase==2){
par(mfrow=c(1,1));
matplot(x=x,y=u1_plot,type="l",xlab="x",ylab="u1(x,t),
        t=0,1,2,3,4,5",xlim=c(xl,xu),lty=1,main="u1(x,t);
        t=0,1,2,3,4,5;",col="black",lwd=2);
par(mfrow=c(1,1));
matplot(x=x,y=u2_plot,type="l",xlab="x",ylab="u2(x,t),
        t=0,1,2,3,4,5",xlim=c(xl,xu),lty=1,main="u2(x,t);
        t=0,1,2,3,4,5;",col="black",lwd=2);
}
#
# Next case
}
```

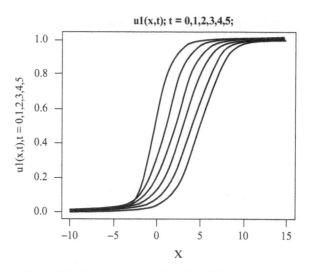

Figure 2.3   Numerical solution of eq. (2.1a), `ncase=2`

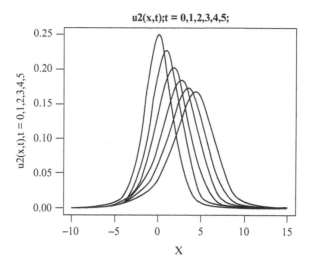

Figure 2.4   Numerical solution of eq. (2.1b), `ncase=2`

The final } concludes the for in ncase.

MOL/ODE function angio_1 called by ode, which has the MOL programming of eqs. (2.1), is considered next.

(2.2.2) MOL/ODE routine

```
angio_1=function(t,u,parms){
#
```

```r
# Function angio_1 computes the t derivative vectors of
# u1(x,t), u2(x,t)
#
# One vector to two vectors
  u1=rep(0,n);u2=rep(0,n);
  for(i in 1:n){
    u1[i]=u[i];
    u2[i]=u[i+n];
  }
#
# u1x, u2x
  u1x=rep(0,n);u2x=rep(0,n);
  u1x=dss004(xl,xu,n,u1);
  u2x=dss004(xl,xu,n,u2);
#
# BCs
  u1x[1]=0; u1x[n]=0;
  u2x[1]=0; u2x[n]=0;
#
# Nonlinear term
  u1u2x=rep(0,n);
  for(i in 1:n){
    u1u2x[i]=chi*u1[i]^(-alpha)*u2[i]*u1x[i];
  }
#
# u1u2xx, u2xx
  u1xx=rep(0,n);u2xx=rep(0,n);u1u2xx=rep(0,n);
  u1xx  =dss004(xl,xu,n,  u1x);
  u2xx  =dss004(xl,xu,n,  u2x);
  u1u2xx=dss004(xl,xu,n,u1u2x);
#
# PDEs
  u1t=rep(0,n);u2t=rep(0,n);
  for(i in 1:n){
    u1t[i]=eps*u1xx[i]-k*u1[i]^m1*u2[i]^m2;
    u2t[i]=d*(u2xx[i]-u1u2xx[i]);
  }
#
# Two vectors to one vector
  ut=rep(0,2*n);
  for(i in 1:n){
    ut[i]   =u1t[i];
    ut[i+n]=u2t[i];
  }
#
# Increment calls to angio_1
```

```
  ncall <<- ncall+1;
#
# Return derivative vector
  return(list(c(ut)));
  }
```

Listing 2.2: MOL/ODE function angio_1 for Eqs. (2.1)

We can note the following details for angio_1.

- The function is defined.

```
  angio_1=function(t,u,parms){
#
# Function angio_1 computes the t derivative vectors of
# u1(x,t), u2(x,t)
```

- The 82-vector of ODE-dependent variables, u, coming into angio_1 as the second argument is placed in two 41-vectors, u1, u2, for the programming of eqs. (2.1).

```
#
# One vector to two vectors
  u1=rep(0,n);u2=rep(0,n);
  for(i in 1:n){
    u1[i]=u[i];
    u2[i]=u[i+n];
  }
```

- The derivatives $u_{1x}(= \frac{\partial u_1}{\partial x})$ and $u_{2x}(= \frac{\partial u_2}{\partial x})$ in eqs. (2.1) are computed by function dss004.

```
#
# u1x, u2x
  u1x=rep(0,n);u2x=rep(0,n);
  u1x=dss004(xl,xu,n,u1);
  u2x=dss004(xl,xu,n,u2);
```

- BCs (2.3) are programmed. Note the use of subscripts 1,n corresponding to $x = -10, 15$ which are in effect BCs at $\pm\infty$ as will be demonstrated in the output.

```
#
# BCs
  u1x[1]=0; u1x[n]=0;
  u2x[1]=0; u2x[n]=0;
```

- The nonlinear chemotaxis diffusion term in eq. (2.1b), $\chi u_1^{-\alpha} u_2 u_{1x}$, is programmed in a for.

```
#
# Nonlinear term
  u1u2x=rep(0,n);
  for(i in 1:n){
    u1u2x[i]=chi*u1[i]^(-alpha)*u2[i]*u1x[i];
  }
```

This coding demonstrates the ease of including nonlinearities in numerical MOL solutions.

- The second derivatives $u_{1xx}, u_{2xx}, (\chi u_1^{-\alpha} u_2 u_{1x})_x$, are computed with second calls to dss004. This is an example of *stagewise* differentiation in which a second-order derivative is computed as the derivative of a first-order derivative.

```
#
# u1u2xx, u2xx
  u1xx=rep(0,n);u2xx=rep(0,n);u1u2xx=rep(0,n);
  u1xx   =dss004(xl,xu,n,   u1x);
  u2xx   =dss004(xl,xu,n,   u2x);
  u1u2xx=dss004(xl,xu,n,u1u2x);
```

- The MOL algorithm for eqs. (2.1) is implemented in a for over the interval $-10 \leq x \leq 15$.

```
#
# PDEs
  u1t=rep(0,n);u2t=rep(0,n);
  for(i in 1:n){
    u1t[i]=eps*u1xx[i]-k*u1[i]^m1*u2[i]^m2;
    u2t[i]=d*(u2xx[i]-u1u2xx[i]);
  }
```

The close resemblance of this coding to eqs. (2.1) demonstrates a principal advantage of the MOL.

- The two 41-vectors with the derivatives in t, u1t, u2t, are placed in a 82-vector, u, to return to the ODE integrator lsodes called in the main program of Listing 2.1.

```
#
# Two vectors to one vector
  ut=rep(0,2*n);
  for(i in 1:n){
    ut[i]   =u1t[i];
    ut[i+n]=u2t[i];
  }
```

- The counter for the calls to angio_1 is incremented and the value returned to the main program of Listing 2.1 by a <<-.

```
#
# Increment calls to angio_1
  ncall <<- ncall+1;
```

- The derivative vector ut is returned for the numerical integration by lsodes using c (the R vector operator), list (lsodes requires a list), and return.

```
#
# Return derivative vector
  return(list(c(ut)));
  }
```

The final{ concludes angio_1.

This concludes the discussion of the MOL/ODE routine angio_1. The subordinate routines for the analytical solution are considered next.

(2.2.3) Subordinate routines

The two subordinate routines, u1_anal, u2_anal, for the analytical solutions of eqs. (2.4) are listed and discussed next.

```
u1_anal=function(x,t,k,d,c){
#
# Function u1_anal computes the analytical solution for u1(x,t)
#
  z=x-c*t;
  u1a=1/(1+exp(-c*z/d));
#
# Return solution
  return(c(u1a));
  }
```

Listing 2.3a: Function u1_anal for the analytical solution of Eq. (2.4a)

```
u2_anal=function(x,t,k,d,c){
#
# Function u2_anal computes the analytical solution for u2(x,t)
#
  z=x-c*t;
  u2a=(c^{2}/(k*d))*exp(-c*z/d)/(1+exp(-c*z/d))^{2};
#
# Return solution
  return(c(u2a));
  }
```

Listing 2.3b: Function u2_anal for the analytical solution of Eq. (2.4b)

Functions u1_anal, u2_anal are straightforward implementations of eqs. (2.4). Note, in particular, the use of the Lagrangian variable $z = x - ct$ for a traveling wave moving at velocity c [1]. This traveling-wave property will be evident when considering the graphical output in Figs. 2.1, 2.2.

(2.3) Model output

The numerical and graphical outputs from the main program of Listing 2.1 are considered next.

(2.3.1) Comparison of numerical and analytical solutions

For ncase=1 in Listing 2.1, the analytical solution of eq. (2.4) is computed and compared with the numerical solution. Abbreviated numerical output is listed in Table 2.1. We can note the following details about this output.

- The numerical value $u_1(x = 5, t = 5) = 0.50001$ is the half-way point for the change in u_1. The analytical value $u_1^a(x = 5, t = 5) = 0.50000$ is the half-way point for the change in u_1^a. The numerical and analytical solutions for u_1 are in close agreement.

```
5.0     5.00    0.50001        0.50000        0.000010
                0.25002        0.25000        0.000018
```

  The numerical value $u_2(x = 5, t = 5) = 0.25002$ is the maximum value for u_2. The analytical value $u_2^a(x = 5, t = 5) = 0.25000$ is the maximum value for u_2^a. The numerical and analytical solutions for u_2 are in close agreement.

- The half-way value of u_1, 0.5, and the maximum value of u_2, 0.25, are in agreement with $z = x - vt = 5 - (1)(5) = 0$ for the traveling-wave solution of eqs. (2.4) (these are the same values as for the ICs of eqs. (2.5), with $z = 0$ so that $z = x - ct = 0 - (1)(0) = 0$).

- The computational effort for the calculation of the complete solution is modest.

```
ncall = 295
```

In summary, the analytical solution of eqs. (2.4) confirms the numerical solution of eqs. (2.1). The traveling-wave solution and the close agreement of the numerical and analytical solutions are evident in Figs. 2.1 and 2.2.

The traveling-wave characteristic[6] of $u_1(x, t)$ is clear. Note also that BCs (2.3) are well approximated since the slope of the numerical solution is close to zero at $x = -10, 15$.

Again, the traveling-wave characteristic of the solution is evident.

Now that the numerical MOL has been validated by comparison with the analytical solution of eqs. (2.4), the PDE system can be studied numerically for cases for which an analytical solution is not available.

---

[6]The use of "characteristic" has a mathematical interpretation since $z = x - ct = constant$ is termed the *characteristic* of the PDE system. In fact, an effective method for the numerical solution of hyperbolic PDEs is termed the *method of characteristics* (MOC). MOC is not used in this book since it requires a specific development for each application that can be rather complicated.

t	x	u1(x,t)	u1_ex(x,t)	u1_err(x,t)
		u2(x,t)	u2_ex(x,t)	u2_err(x,t)
5.0	-10.00	0.00000	0.00000	0.000001
		0.00000	0.00000	0.000003
5.0	-9.75	0.00000	0.00000	0.000001
		0.00000	0.00000	0.000003

.
.
.

Output for x = -9.50 to x = 3.75 removed

.
.
.

t	x	u1(x,t)	u1_ex(x,t)	u1_err(x,t)
5.0	4.00	0.26894	0.26894	-0.000003
		0.19657	0.19661	-0.000044
5.0	4.25	0.32081	0.32082	-0.000007
		0.21787	0.21789	-0.000029
5.0	4.50	0.37754	0.37754	-0.000006
		0.23500	0.23500	-0.000009
5.0	4.75	0.43782	0.43782	0.000001
		0.24614	0.24613	0.000009
5.0	5.00	0.50001	0.50000	0.000010
		0.25002	0.25000	0.000018
5.0	5.25	0.56220	0.56218	0.000019
		0.24615	0.24613	0.000013
5.0	5.50	0.62248	0.62246	0.000024
		0.23500	0.23500	0.000000
5.0	5.75	0.67920	0.67918	0.000024
		0.21788	0.21789	-0.000015
5.0	6.00	0.73108	0.73106	0.000019
		0.19659	0.19661	-0.000025

.
.
.

Output for x = 6.25 to x = 14.50 removed

.
.
.

t	x	u1(x,t)	u1_ex(x,t)	u1_err(x,t)
5.0	14.75	0.99991	0.99994	-0.000035
		0.00009	0.00006	0.000035
5.0	15.00	0.99991	0.99995	-0.000044
		0.00009	0.00005	0.000045

ncall =   295

Table 2.1: Abbreviated Numerical Output at $t = 5$ for ncase=1

(2.3.2) Effect of diffusion on the traveling-wave solution

As a parameter variation, ϵ in eq. (2.1a) is changed from $\epsilon = 0$ for ncase=1 to $\epsilon = 1$ for ncase2 in the main program of Listing 2.1.

```
#
# Parameters
  c=1;d=1;k=1;m1=0;m2=1;chi=2;alpha=1;
  if(ncase==1){eps=0;}
  if(ncase==2){eps=1;}
```

This in effect adds diffusion to eq. (2.1a). As we can observe in Figs. 2.3 and 2.4, the change in ϵ has a major effect on the solution.

We can note the following points about Figs. 2.3, 2.4.

- The solution for $u_1(x, t)$ in Fig. 2.3 is no longer a traveling wave. For example, the individual curves are not equally spaced as they are in Fig. 2.1 (which follows for a solution that is a function of $z = x - ct$ only, for equally spaced t ($t = 0, 1, 2, 3, 4, 5$)).
- In Fig. 2.4, the solution for $u_2(x, t)$ is no longer a function of z only as in eq. (2.4b). That is, the shape of the function changes with increasing t and the diffusion from $\epsilon = 1$ has the expected effect of smoothing the solution.

The routines in Listings 2.1, 2.2, and 2.3 can be used for a variety of numerical experiments, both with regard to parameter changes and modification of the PDEs.

(2.4) Conclusions

In the preceding discussion, a physical interpretation of the RHS terms in eqs. (2.1) has not been included to conserve space. Also, this type of interpretation is available from an extensive literature on chemotaxis and angiogenesis. Rather, the emphasis has been on numerical methods, and associated routines, for the solution of eqs. (2.1) that can then be applied to other PDE systems.

Among the possibilities for further study are

- Numerical examination of the RHS terms of eqs. (2.1) and their relative contribution to the LHS derivatives in t. This is readily accomplished since $u_1(x, t)$ and $u_2(x, t)$ are available to compute and display these terms. Examples of this type of analysis are given in [4]. Generally, this type of examination of the PDE RHS terms provides an explanation of the principal features of the solution as defined by the LHS derivatives in t. This in turn facilitates experimentation with the model such as modification of the RHS terms and addition of new RHS terms.
- Use of other coordinates, such as spherical coordinates, to match the physical geometry of interest, e.g., a tumor. This change in coordinates would be straightforward if the PDEs are to remain 1D. For example, the Cartesian coordinate x in

the Fickian diffusion term of eq. (2.1a)

$$u_{1xx} = \frac{\partial^2 u_1}{\partial x^2}$$

would be replaced with the spherical radial coordinate r

$$\frac{1}{r^2}\frac{\partial \left(r^2 \frac{\partial u_1}{\partial r} \right)}{\partial r} = \frac{1}{r^2}(r^2 u_{1r})_r$$

An aspect of this change of variables not encountered with the preceding analysis in x is the use of the derivative in r at $r = 0$ (regularizing the singularity in r at $r = 0$, which can usually be done with *l'Hospital's rule*).

- Changing the ICs (rather than using eqs. (2.5)) to reflect the initial distributions of endothelial cells and VEGF. This would be done for `ncase` ≥ 2 if the analytical solution is to be retained for `ncase = 1`.

These possibilities for numerical experimentation demonstrate the utility of the numerical approach to the PDE model since equivalent analytical experimentation would be difficult. The library routines, e.g., `ode`, `dss004`, can be used for convenience and with reliability for further development. The general framework of the main program in Listing 2.1 and the MOL/ODE routine, `angio_1`, in Listing 2.2 can be a starting point for new PDE model applications. This approach is demonstrated in subsequent chapters.

References

[1] Griffiths, G.W. and W.E. Schiesser (2012), *Traveling Wave Analysis of Partial Differential Equations*, Elsevier/Academic Press, Boston, MA.

[2] Murray, J.D. (2003), *Mathematical Biology, II: Spatial Models and Biomedical Applications*, 3rd edn, Springer-Verlag, New York.

[3] National Cancer Institute information about angiogenesis. http://www.cancer.gov/cancertopics/factsheet/Therapy/angiogenesis-inhibitors; http://oncotherapy.us/ECM/ECM_Angiogenesis_NCI.pdf.

[4] Schiesser, W.E. (2014), *Differential Equation Analysis in Biomedical Science and Engineering: Partial Differential Equation Applications in R*, John Wiley & Sons, Inc., Hoboken, NJ.

[5] Wang, Z.-A. (2012), Wavefront of an angiogenesis model, *Discrete and Continuous Dynamical Systems, Series B*, **17**, 8, 2849–2860.

3

THERMOGRAPHIC TUMOR LOCATION

The mathematical model discussed in this chapter describes heat conduction in the tissue resulting from heat release by a tumor. This heat release results in an elevated temperature on the skin surface near the tumor. This pattern (image) can be detected by an infrared sensor, and this procedure, termed *thermography*, can therefore be used to detect the tumor. The following statements [1] give some additional background.

> In non-invasive thermal diagnostics, accurate correlations between the thermal image on skin surface and interior human physiology are often desired, which require general solutions for the bioheat equation.

> In general the body surface temperature is controlled by the blood circulation underneath the skin, local metabolism, and the heat exchange between the skin and its environment. Changes in any of these parameters can induce variations of temperature and heat flux at the skin surface reflecting the physiologic state of the human body. The particular tumor architecture and angiogenesis processes can lead to an abnormal situation. Inflammation, metabolic rate, interstitial hypertension, abnormal vessel morphology and lack of response to homeostatic signals are some of the particular features that make tumors behave differently than normal tissue in terms of heat production and dissipation. Temperatures at skin above a breast tumor or a malignant melanoma, a tumor of melanocytes, which are found predominantly in skin, have been found to be several degrees higher than that of the surrounding area. So, the abnormal temperature at skin surface can be used in order to predict the location, size and thermal properties of the tumor region as well as to study the tumor evolution after a treatment procedure.

Method of Lines PDE Analysis in Biomedical Science and Engineering, First Edition. William E. Schiesser.
© 2016 John Wiley & Sons, Inc. Published 2016 by John Wiley & Sons, Inc.
Companion website: www.wiley.com/go/Schiesser/PDE_Analysis

As suggested in these statements, the objectives of this chapter are to

- Develop a 2D 1-PDE model based on the *Pennes bioheat equation* [1,2] in cylindrical coordinates, including the initial condition (IC) and boundary conditions (BCs).
- Code (program) a method of lines (MOL) solution of the model to give temperature profiles throughout the tissue that includes the tumor and the skin surface.
- Investigate details and features of the model, such as the
  - Tumor volumetric heat generation rate.
  - Effect on the skin surface image of the tumor depth.
  - Effect on the model solution of heat transfer parameters such as the air-to-skin heat transfer coefficient and the perfusion rate that determines the blood temperature regulation of the tissue.
  - Changes in the size of the tumor.
  - Spatial gridding to achieve acceptable accuracy (convergence) of the numerical solution.

The mathematical model for thermographic tumor detection is formulated next. A detailed derivation of a 3D PDE in cylindrical coordinates is given as additional background in Appendix A.

(3.1) 2D, 1-PDE model

The PDE model is formulated in cylindrical coordinates, (r, θ, z), starting with the heat conduction equation (Fourier's second law), which is a special case of eq. (A.5), Appendix A, with the dependent variable taken as $T(r, z, t)$. [1]

$$C\frac{\partial T}{\partial t} = \frac{1}{r}\frac{\partial}{\partial r}\left(k_r \frac{\partial T}{\partial r}\right) + \frac{\partial}{\partial z}\left(k_z \frac{\partial T}{\partial z}\right) \tag{3.1a}$$

where

Variable, parameter	Interpretation
T	tissue temperature (°C)
t	time (s)
r	radial position in the thermal domain (cm)
z	axial position in the thermal domain (cm)
C	tissue volumetric heat capacity (J/cm^3 °C)
k_r, k_z	tissue thermal conductivity in r, z directions (J cm/s cm^2 °C)

---

[1] The dependent variable, tissue temperature, is denoted as $T(r, z, t)$ (for heat transfer) rather than $c(r, z, t)$ as in Appendix A (for mass transfer). Also, this is a departure from the convention of designating the PDE dependent variable as $u_1(r, z, t)$, primarily because of the use of T in eqs. (3.1c), (3.2), (3.3), e.g., T_b, T_0, T_a.

Variation of T with θ is neglected so that eq. (3.1a) has two spatial independent variables, (r, z).

Eq. (3.1a) has a variable coefficient that forms a singularity ($1/r$ for $r = 0$), and nonlinear coefficients resulting from temperature-dependent physical properties ($C(T)$, $k_r(T)$, $k_z(T)$). In the following analysis, temperature-independent (constant) properties are assumed[2] so that eq. (3.1a) becomes

$$C\frac{\partial T}{\partial t} = k_r \frac{1}{r}\frac{\partial}{\partial r}\left(\frac{\partial T}{\partial r}\right) + k_z\frac{\partial}{\partial z}\left(\frac{\partial T}{\partial z}\right)$$

or after expanding the RHS derivative in r

$$C\frac{\partial T}{\partial t} = k_r\left(\frac{\partial^2 T}{\partial r^2} + \frac{1}{r}\frac{\partial T}{\partial r}\right) + k_z\left(\frac{\partial^2 T}{\partial z^2}\right) \tag{3.1b}$$

The tissue is assumed to be *isotropic* (the physical properties are spatially uniform or direction independent) so that $k_r = k_z = k$.

We next include a term for heat transfer between the tissue and blood that can be at different temperatures, T and T_b, respectively. Also, a volumetric heat source term is added for the tumor, $Q_m(r, z, t)$.

$$C\frac{\partial T}{\partial t} = k\left(\frac{\partial^2 T}{\partial r^2} + \frac{1}{r}\frac{\partial T}{\partial r}\right) + k\left(\frac{\partial^2 T}{\partial z^2}\right) + k_e(T_b - T) + Q_m \tag{3.1c}$$

where

Variable, parameter	Interpretation
T_b	blood temperature (°C)
$k_e = G_b C_b$	perfusion coefficient (J/s cm³ °C)
G_b	blood perfusion rate (s⁻¹)
C_b	blood volumetric specific heat (J/cm³ °C)
Q_m	volumetric heat source (J/cm³ s)

Eq. (3.1c), termed the *Pennes bioheat transfer equation* [1], is the starting point for the MOL analysis that follows. The numerical solution to eq. (3.1c) gives the tissue temperature, $T(r, z, t)$, throughout the spatial domain $0 \le r \le r_0$, $0 \le z \le z_L$, where r_0 and z_L are the dimensions of the cylindrical spatial domain.

Eq. (3.1c) is first order in t and second order in r and z. It therefore requires one initial condition (IC) in t and two boundary conditions (BCs) in r and z. We take these auxiliary conditions as

$$T(r, z, t = 0) = T_0 \tag{3.2}$$

$$\frac{\partial T(r = 0, z, t)}{\partial r} = 0; \quad \frac{\partial T(r = r_0, t)}{\partial r} = 0 \tag{3.3a,b}$$

---

[2]Temperature-dependent tissue properties can be handled numerically as illustrated in [4], Chapter 8.

$$-k\frac{\partial T(r, z = 0, t)}{\partial z} = h(T_a - T(r, z = 0, t)); \quad \frac{\partial T(r, z = z_L, t)}{\partial z} = 0 \quad (3.3c,d)$$

where

Parameter	Interpretation
T_0	tissue initial (normal) temperature (°C)
r_0	radius of the thermal domain (cm)
z_L	length of the thermal domain (cm)
T_a	ambient temperature (°C)
h	skin surface to ambient heat transfer coefficient (J/s cm^2 °C)

BCs (3.3a,b,d) are termed *Neumann* since the first derivatives of $T(r, z, t)$ are specified. BC (3.3c) is termed *third type* or *Robin* since it includes the dependent variable $T(r, z = 0, t)$ and its derivative in z, $\dfrac{\partial T(r, z = 0, t)}{\partial z}$.

Eqs. (3.2) and (3.3) have the following interpretation.

- IC (3.2): The tissue temperature starts at the normal temperature T_0.
- BC (3.3a): Symmetry condition at the centerline $r = 0$.
- BC (3.3b): r_0 is large enough that there is no radial variation at $r = r_0$.
- BC (3.3c): The skin-to-air heat transfer rate is proportional to the temperature difference $T_a - T(r, z = 0, t)$.
- BC (3.3d): The axial dimension is large enough that there is no axial variation at $z = z_L$.

The method of lines (MOL) programming of the model consisting of eqs. (3.1c), (3.2), (3.3), is considered next.

(3.2) MOL analysis

Eq. (3.1c) is approximated as a system of ordinary differential equations (ODEs) within the MOL context. The programming for the numerical integration of this MOL/ODE system is considered next.

(3.2.1) ODE routine

The programming of eqs. (3.1c) and (3.3) is in the following routine.

```
  thermo_1=function(t,u,parms){
#
# Function thermo_1 computes the t derivative vector
# of T(r,z,t)
#
```

```
# 1D vector to 2D matrix
   T=matrix(0,nrow=nr,ncol=nz);
   Tt=matrix(0,nrow=nr,ncol=nz);
   for(i in 1:nr){
   for(j in 1:nz){
     T[i,j]=u[(i-1)*nz+j];
   }
   }
#
# BCs
   Tr=matrix(0,nrow=nr,ncol=nz);
   Tz=matrix(0,nrow=nr,ncol=nz);
#
#   r=0
     Tr[1,]=0;
#
#   r=r0
     Tr[nr,]=0;
#
#   z=0
     Tz[,1]=-(h/k)*(Ta-T[,1]);
#
#   z=zL
     Tz[,nz]=0;
#
# Trr
     Trr=matrix(0,nrow=nr,ncol=nz);
     nl=2;nu=2;
     for(j in 1:nz){
       Trr[,j]=dss044(0,r0,nr,T[,j],Tr[,j],nl,nu);
     }
#
# Tzz
     Tzz=matrix(0,nrow=nr,ncol=nz);
     nl=2;nu=2;
     for(i in 1:nr){
       Tzz[i,]=dss044(0,zL,nz,T[i,],Tz[i,],nl,nu);
     }
#
# MOL/ODEs at grid points in r and z
   for(i in 1:nr){
   for(j in 1:nz){
     if(i==1){
       Tt[i,j]=(1/C)*k*2*Trr[i,j]+k*Tzz[i,j]+
                 ke*(Tb-T[i,j])+Qm[i,j];
     }else{
```

```
        Tt[i,j]=(1/C)*k*(Trr[i,j]+1/r[i]*Tr[i,j])+k*Tzz[i,j]+
                  ke*(Tb-T[i,j])+Qm[i,j];
      }
#
# Next j
  }
#
# Next i
  }
#
# 2D matrix to 1D vector
  ut=rep(0,nr*nz);
  for(i in 1:nr){
  for(j in 1:nz){
    ut[(i-1)*nz+j]=Tt[i,j];
  }
  }
#
# Increment calls to thermo_1
  ncall <<- ncall+1;
#
# Return derivative vector
  return(list(c(ut)));
}
```

Listing 3.1: MOL/ODE routine for Eqs. (3.1c) and (3.3)

We can note the following details about Listing 3.1.

- The function is defined.

  ```
  thermo_1=function(t,u,parms){
  #
  # Function thermo_1 computes the t derivative vector
  # of T(r,z,t)
  ```

- The input vector u is placed in a 2D matrix T to facilitate the programming of eqs. (3.1c), (3.2), and (3.3). The first (row) and second (column) subscripts pertain to r and z, respectively, in eq. (3.1c).

  ```
  #
  # 1D vector to 2D matrix
    T=matrix(0,nrow=nr,ncol=nz);
    Tt=matrix(0,nrow=nr,ncol=nz);
    for(i in 1:nr){
    for(j in 1:nz){
  ```

```
    T[i,j]=u[(i-1)*nz+j];
  }
}
```

- BCs (3.3) are programmed. First, matrices for $\dfrac{\partial T}{\partial r}$ and $\dfrac{\partial T}{\partial z}$ defined at the boundaries are declared.

```
#
# BCs
  Tr=matrix(0,nrow=nr,ncol=nz);
  Tz=matrix(0,nrow=nr,ncol=nz);
```

The four BCs, eqs. (3.3), are then programmed.
  – BC (3.3a) with the first subscript 1 for $r = 0$.

```
#
#   r=0
  Tr[1,]=0;
```

  – BC (3.3b) with the first subscript nr for $r = r_0$.

```
#
#   r=r0
  Tr[nr,]=0;
```

  – BC (3.3c) with the second subscript 1 for $z = 0$.

```
#
#   z=0
  Tz[,1]=-(h/k)*(Ta-T[,1]);
```

  – BC (3.3d) with the second subscript nz for $z = z_L$.

```
#
#   z=zL
  Tz[,nz]=0;
```

A vectorized operator is used, e.g., Tr[1,] for all values of the second subscript corresponding to all values of z with $r = 0$.

- The second derivative $\dfrac{\partial^2 T}{\partial r^2}$, Trr, in eq. (3.1c) is computed by dss044.

```
#
# Trr
    Trr=matrix(0,nrow=nr,ncol=nz);
    nl=2;nu=2;
```

```
for(j in 1:nz){
  Trr[,j]=dss044(0,r0,nr,T[,j],Tr[,j],nl,nu);
}
```

The for steps through the values of z, $0 \leq z \leq z_L$. The vector operator in Trr[,j] includes all values of r for $0 \leq r \leq r_0$. Briefly, the input arguments of dss044 are

- 0,r0,nr: Boundary values of r defining the interval in r, $0 \leq r \leq r_0$, over which the second derivative Trr is computed on a grid of nr points.
- T[,j]: Vector of $T(r,z,t)$ values over r at a specific z (point j) for which the second derivative Trr in computed.
- Tr[,j]: Vector of first derivatives in r, including the boundary values defined previously.
- nl,nu: Specification of the type of BCs, in this case nl=2,nu=2, for specification of the first derivative in r at the boundaries, that is, for Neumann BCs (3.3a), (3.3b). nl=1,nu=1 (not used) would specify T at the boundaries, that is, Dirichlet BCs.

The dss (Differentiation in Space Subroutines) are discussed in detail in [3].

- The second derivative $\dfrac{\partial^2 T}{\partial z^2}$, Tzz, in eq. (3.1c) is computed by dss044.

```
#
# Tzz
    Tzz=matrix(0,nrow=nr,ncol=nz);
    nl=2;nu=2;
    for(i in 1:nr){
      Tzz[i,]=dss044(0,zL,nz,T[i,],Tz[i,],nl,nu);
    }
```

The programming is analogous to that for Trr. BC (3.3c) specifies the derivative at the boundary $z = 0$ so that nl=2, and BC (3.3d) specifies the derivative at $z = z_L$ so that nu=2.

- All of the spatial (boundary value) derivatives in eq. (3.1c) have been computed, so $\dfrac{\partial T}{\partial t}$ of eq. (3.1c) is programmed (within the MOL format).

```
#
# MOL/ODEs at grid points in r and z
    for(i in 1:nr){
    for(j in 1:nz){
      if(i==1){
        Tt[i,j]=(1/C)*k*2*Trr[i,j]+k*Tzz[i,j]+
                ke*(Tb-T[i,j])+Qm[i,j];
      }else{
        Tt[i,j]=(1/C)*k*(Trr[i,j]+1/r[i]*Tr[i,j])+k*Tzz[i,j]+
                ke*(Tb-T[i,j])+Qm[i,j];
      }
```

```
#
# Next j
  }
#
# Next i
  }
```

Note the two nested for's to cover the 2D domain $(0 \leq r \leq r_0) \times (0 \leq z \leq z_L)$. At $r = 0$ (i=1), the first derivative term in r in eq. (3.1c) is indeterminate $(0/0)$. The singularity $1/r$ can be regularized using *l'Hospital's rule* and BC (3.3a).

$$\lim_{r \to 0} \frac{1}{r} \frac{\partial T}{\partial r} = \frac{\partial^2 T}{\partial r^2} \tag{3.4}$$

This result is programmed for i=1, that is,

$$\left(\frac{\partial^2 T}{\partial r^2} + \frac{1}{r} \frac{\partial T}{\partial r} \right)_{r=0} = 2 \frac{\partial^2 T}{\partial r^2}$$

or 2*Trr[i,j] within the inner for.

- All of the values of $\dfrac{\partial T}{\partial t}$ in eq. (3.1c) are now defined numerically and are then placed in a vector ut to return to the ODE integrator called in the main program (considered next).

```
#
# 2D matrix to 1D vector
  ut=rep(0,nr*nz);
  for(i in 1:nr){
  for(j in 1:nz){
    ut[(i-1)*nz+j]=Tt[i,j];
  }
  }
```

- The number of calls to thermo_1 is incremented and returned to the main program with <--.

```
#
# Increment calls to thermo_1
  ncall <<- ncall+1;
```

- The derivative vector ut is returned to the ODE integrator with a combination of c (the vector operator in R), list (the ODE integrator requires a list for the derivative vector), and return.

```
#
# Return derivative vector
```

```
    return(list(c(ut)));
  }
```

The final } concludes function thermo_1.

The main program that calls thermo_1 of Listing 3.1 is next.

(3.2.2) Main program

The main program that calls ODE integrator lsodes is in Listing 3.2.

```
#
# Thermographic tumor detection
#
# Access ODE integrator
  library("deSolve");
#
# Access functions for numerical solutions
  setwd("g:/chap3");
  source("thermo_1.R");
  source("dss044.R");
#
# Model parameters
  C=3; k=0.002; h=0.0005; ke=0.001; r0=1; zL=2;
  Ta=(70-32)/1.8;T0=(98.6-32.0)/1.8;Tb=T0;
#
# Radial grid
  nr=11;
  r=seq(from=0,to=r0,by=(r0-0)/(nr-1));
#
# Axial grid
  nz=11;
  z=seq(from=0,to=zL,by=(zL-0)/(nz-1));
#
# Heat source
  Qm=matrix(0,ncol=nr,nrow=nz);
  nrs=1;nzs=4;
  Qm[nrs,nzs]=1;
#
# Grid in t
  nout=4;tf=6*60;
  tout=seq(from=0,to=tf,by=tf/(nout-1));
#
# Display selected parameters
  cat(sprintf(
    "\n\n   nr = %2d   nz = %2d\n",nr,nz));
```

```
#
# Initial condition
  u0=rep(0,nr*nz);
  for(i in 1:nr){
  for(j in 1:nz){
    u0[(i-1)*nz+j]=T0;
  }
  }
  ncall=0;
#
# ODE integration
  out=lsodes(func=thermo_1,y=u0,times=tout,parms=NULL);
  nrow(out)
  ncol(out)
#
# Calls to ODE routine
  cat(sprintf("\n\n    ncall = %5d\n\n",ncall));
#
# Vector to matrix
  T =matrix(0,nrow=nr,ncol=nz);
  T1=matrix(0,nrow=nr,ncol=nz);
  T2=matrix(0,nrow=nr,ncol=nz);
  T3=matrix(0,nrow=nr,ncol=nz);
  T4=matrix(0,nrow=nr,ncol=nz);
  for(it in 1:nout){
    for(i in 1:nr){
    for(j in 1:nz){
      T[i,j]=out[it,(i-1)*nr+j+1];
      cat(sprintf("\n t = %3.0f   r = %4.2f  z = %4.2f
                  T = %5.2f",tout[it],r[i],z[j],T[i,j]));
    }
    }
    cat(sprintf("\n"));
    if(it==1){T1=T;}
    if(it==2){T2=T;}
    if(it==3){T3=T;}
    if(it==4){T4=T;}
  }
#
# 2 x 2 array of plots
  par(mfrow=c(2,2));
  matplot(x=r,y=T1,type="l",xlab="r",ylab="T(r,z,t=0)",
          xlim=c(0,r[nr]),ylim=c(34,39),lty=1,
          main="T(r,z,t=0),\n z=0,0.2,...,2",lwd=2);
  matplot(x=r,y=T2,type="l",xlab="r",ylab="T(r,z,t=120)",
          xlim=c(0,r[nr]),ylim=c(34,39),lty=1,
```

```
        main="T(r,z,t=120),\n z=0,0.2,...,2",lwd=2);
matplot(x=r,y=T3,type="l",xlab="r",ylab="T(r,z,t=240)",
        xlim=c(0,r[nr]),ylim=c(34,39),lty=1,
        main="T(r,z,t=240),\n z=0,0.2,...,2",lwd=2);
matplot(x=r,y=T4,type="l",xlab="r",ylab="T(r,z,t=360)",
        xlim=c(0,r[nr]),ylim=c(34,39),lty=1,
        main="T(r,z,t=360),\n z=0,0.2,...,2",lwd=2);
```

Listing 3.2: Main program for Eqs. (3.1c), (3.2), and (3.3)

We can note the following details about Listing 3.2.

- The ODE integrator library deSolve and the subordinate routines thermo_1, dss044 are accessed. The setwd (set working directory) has to be edited for the local computer (note the use of / rather than the usual \).

```
#
# Access ODE integrator
  library("deSolve");
#
# Access functions for numerical solutions
  setwd("c:/chap3");
  source("thermo_1.R");
  source("dss044.R");
```

- The parameters of eqs. (3.1c), (3.2), and (3.3) are defined numerically.

```
#
# Model parameters
  C=3; k=0.002; h=0.0005; ke=0.001; r0=1; zL=2;
  Ta=(70-32)/1.8;T0=(98.6-32.0)/1.8;Tb=T0;
```

cgs units are used for these parameters, e.g.,

Parameter	Name	Units
C	volumetric specific heat	$J/cm^3 \,^{\circ}C$
k	thermal conductivity	$J \, cm/s \, cm^2 \,^{\circ}C$
h	skin-to-air heat transfer coefficient	$J/s \, cm^2 \,^{\circ}C$

The other parameters ke,r0,zL,Ta,T0,Tb were discussed previously as part of the derivation of eq. (3.1c). The temperatures are converted from °F to °C.

- A radial grid of 11 points is defined for the interval $0 \leq r \leq r_0$ (in cm).

```
#
# Radial grid
  nr=11;
  r=seq(from=0,to=r0,by=(r0-0)/(nr-1));
```

- An axial grid of 11 points is defined for the interval $0 \leq z \leq z_L$ (in cm).

```
#
# Axial grid
  nz=11;
  z=seq(from=0,to=zL,by=(zL-0)/(nz-1));
```

- A volumetric heat source for the tumor that is in eq. (3.1c), Q_m, is defined at the point $r = 0$, $z = (4 - 1)/(11 - 1)(z_L - 0) = (3/10)(2) = 0.6$ cm from the skin surface (at $z = 0$).

```
#
# Heat source
  Qm=matrix(0,ncol=nr,nrow=nz);
  nrs=1;nzs=4;
  Qm[nrs,nzs]=1;
```

This placement of the tumor heat source can be varied both with respect to the position, e.g., $(r = 0, z = 0.6)$, and the size (using more than one point in r and z). Thus, the effect of tumor position and size on the skin temperature pattern (image) can be studied.

- A grid in t of 4 output points is defined for $t = 0, 120, 240, 360$ (s).

```
#
# Grid in t
  nout=4;tf=6*60;
  tout=seq(from=0,to=tf,by=tf/(nout-1));
```

- The grid dimensions of the thermal domain are displayed.

```
#
# Display selected parameters
  cat(sprintf(
    "\n\n    nr = %2d    nz = %2d\n",nr,nz));
```

- The IC of eq. (3.2) is a constant over the thermal domain, T0 $= 98.6$ °F (previously defined and converted to °C).

```
#
# Initial condition
  u0=rep(0,nr*nz);
  for(i in 1:nr){
  for(j in 1:nz){
    u0[(i-1)*nz+j]=T0;
  }
  }
  ncall=0;
```

This IC is placed in a vector, u0, that is, an input to the ODE integrator lsodes (considered next). The length of u0 ($= 11 \times 11 = 121$) informs lsodes of the number of ODEs to be integrated. Finally, the counter for the calls to thermo_1 is initialized.

- The 121-ODE vector is integrated by lsodes.

```
#
# ODE integration
  out=lsodes(func=thermo_1,y=u0,times=tout,parms=NULL);
  nrow(out)
  ncol(out)
```

The input arguments to lsodes are: (1) the MOL/ODE function, thermo_1, of Listing 3.1; (2) the IC vector, u0; (3) the vector of output values of t, tout; and (4) parameters passed to thermo_1, parms, (unused in this case); func,y,times,parms are reserved names for the input arguments of lsodes. The numerical ODE solutions are returned in out.

Finally, the row and column dimensions of out, 4,121+1 respectively, are displayed by nrow,nout for confirmation. The additional 1 in the second dimension is used for the value of t (in out(4,1) so that the ODE solutions are in out(4,2) to out(4,122)).

- The number of calls to thermo_1 is displayed to give a measure of the computational effort required to compute the solution.

```
#
# Calls to ODE routine
  cat(sprintf("\n\n   ncall = %5d\n\n",ncall));
```

- Arrays are declared for plotting the numerical solution at $t = 0, 120, 240, 360$.

```
#
# Vector to matrix
  T =matrix(0,nrow=nr,ncol=nz);
  T1=matrix(0,nrow=nr,ncol=nz);
  T2=matrix(0,nrow=nr,ncol=nz);
  T3=matrix(0,nrow=nr,ncol=nz);
  T4=matrix(0,nrow=nr,ncol=nz);
```

- The numerical solution of eq. (3.1c), $T(r, z, t)$, is placed in matrix T using three nested for's: (1) for in it for t, (2) for in nr for r, and (3) for in nz for z. The offset of 1 in (i-1)*nr+j+1 reflects the values of t also placed in out (so the second dimension of out is 122 rather than (11)(11)=121 as explained previously, and the value 1 is for t while 2–122 are for the 121 values of $T(r, z, t)$).

```
for(it in 1:nout){
    for(i in 1:nr){
```

```
      for(j in 1:nz){
        T[i,j]=out[it,(i-1)*nr+j+1];
        cat(sprintf("\n t = %3.0f    r = %4.2f    z = %4.2f
                    T = %5.2f",tout[it],r[i],z[j],T[i,j]));
      }
      }
    cat(sprintf("\n"));
    if(it==1){T1=T;}
    if(it==2){T2=T;}
    if(it==3){T3=T;}
    if(it==4){T4=T;}
    }
```

After the detailed output from the sprintf, the numerical solution is placed in the four matrices T1 to T4 for $t = 0, 120, 240, 360$.

- A 2×2 array of plots is produced with the matplot utility corresponding to $t = 0, 120, 240, 360$. Arrays T1, T2, T3, T4 demonstrate the t evolution of the solution to eqs. (3.1c), (3.2) and (3.3).

```
#
# 2 x 2 array of plots
  par(mfrow=c(2,2));
  matplot(x=r,y=T1,type="l",xlab="r",ylab="T(r,z,t=0)",
          xlim=c(0,r[nr]),ylim=c(34,39),lty=1,
          main="T(r,z,t=0),\n z=0,0.2,...,2",lwd=2);
  matplot(x=r,y=T2,type="l",xlab="r",ylab="T(r,z,t=120)",
          xlim=c(0,r[nr]),ylim=c(34,39),lty=1,
          main="T(r,z,t=120),\n z=0,0.2,...,2",lwd=2);
  matplot(x=r,y=T3,type="l",xlab="r",ylab="T(r,z,t=240)",
          xlim=c(0,r[nr]),ylim=c(34,39),lty=1,
          main="T(r,z,t=240),\n z=0,0.2,...,2",lwd=2);
  matplot(x=r,y=T4,type="l",xlab="r",ylab="T(r,z,t=360)",
          xlim=c(0,r[nr]),ylim=c(34,39),lty=1,
          main="T(r,z,t=360),\n z=0,0.2,...,2",lwd=2);
```

The numerical and graphical outputs are given in Table 3.1 and Fig. 3.1.

The library differentiator, dss044 (used in thermo_1 of Listing 3.1), is discussed in detail in [3]. This completes the programming of eqs. (3.1c), (3.2), and (3.3).

(3.3) Model output

The numerical and graphical outputs from Listings 3.1 and 3.2 in Table 3.1 and Fig. 3.1 are discussed next.

We can note the following details of the abbreviated output in Table 3.1 and the graphical output in Fig. 3.1.

```
nr = 11   nz = 11
nrow(out)
[1] 4
ncol(out)
[1] 122
ncall =    286
t =   0   r = 0.00   z = 0.00   T = 37.00
t =   0   r = 0.00   z = 0.20   T = 37.00
t =   0   r = 0.00   z = 0.40   T = 37.00
t =   0   r = 0.00   z = 0.60   T = 37.00
t =   0   r = 0.00   z = 0.80   T = 37.00
t =   0   r = 0.00   z = 1.00   T = 37.00
                 .          .
                 .          .
                 .          .
      Output for t = 0, r = 0, z = 1.2
          to r = 1, z = 0.8 removed
                 .          .
                 .          .
                 .          .
t =   0   r = 1.00   z = 1.00   T = 37.00
t =   0   r = 1.00   z = 1.20   T = 37.00
t =   0   r = 1.00   z = 1.40   T = 37.00
t =   0   r = 1.00   z = 1.60   T = 37.00
t =   0   r = 1.00   z = 1.80   T = 37.00
t =   0   r = 1.00   z = 2.00   T = 37.00
                 .          .
                 .          .
                 .          .
      Output for t = 120, 240 removed
                 .          .
                 .          .
                 .          .
t = 360   r = 0.00   z = 0.00   T = 37.32
t = 360   r = 0.00   z = 0.20   T = 37.70
t = 360   r = 0.00   z = 0.40   T = 37.97
t = 360   r = 0.00   z = 0.60   T = 39.37
t = 360   r = 0.00   z = 0.80   T = 37.95
t = 360   r = 0.00   z = 1.00   T = 37.54
                 .          .
                 .          .
                 .          .
```

Table 3.1: Abbreviated Output from Listings 3.1 and 3.2

```
     Output for t = 360, r = 0, z = 1.2
         to r = 1, z = 0.8 removed
                    .              .
                    .              .
                    .              .
t = 360    r = 1.00    z = 1.00    T = 36.77
t = 360    r = 1.00    z = 1.20    T = 36.85
t = 360    r = 1.00    z = 1.40    T = 36.91
t = 360    r = 1.00    z = 1.60    T = 36.94
t = 360    r = 1.00    z = 1.80    T = 36.96
t = 360    r = 1.00    z = 2.00    T = 36.97
```

Table 3.1: (*Continued*)

- The grid dimensions of the thermal domain are confirmed.

    ```
    nr = 11    nz = 11
    ```

- The dimensions of the solution matrix out are confirmed.

    ```
    nrow(out)
    [1] 4
    ncol(out)
    [1] 122
    ```

The second dimension of out is $(11)(11) + 1 = 122$ as discussed previously.

- IC (3.2) is confirmed ($T(r, z, t = 0) = 98.6$ °C $= 37.0$ °C) as expected. Checking the IC is worthwhile since if it is incorrect, the remaining solution will be incorrect.
- The computational effort to compute a solution is quite modest, ncall = 286.
- At the centerline $r = 0$, the temperature $T(r = 0, z = 0.60, t = 120) = 39.37$ is a maximum as expected since this is the point where the tumor is located. The maximum temperature rise above the IC, $39.37 - 37.00 = 2.37$ °C, is appreciable, but as observed in the subsequent discussion, it is not transmitted to the skin surface where the temperature image is measured (observed).
- At $t = 240$, the corresponding output (not in Table 3.1) is

    ```
    t = 240    r = 0.00    z = 0.60    T = 39.35
    ```

Thus, the maximum temperature is nearly the same as for $t = 360$ (39.37 vs 39.35) so that the thermal region appears to have reached a nearly *steady state* or *equilibrium* condition (with no further change with respect to t). This conclusion is confirmed by Fig. 3.1. This steady state comes about when the heat generation from the tumor is matched by the heat loss through the skin at $z = 0$ (there is no heat loss through the surfaces at $r = r_0$ and $z = z_L$ since eqs. (3.3b) and (3.3d) are *zero heat flux* or *insulated* BCs).

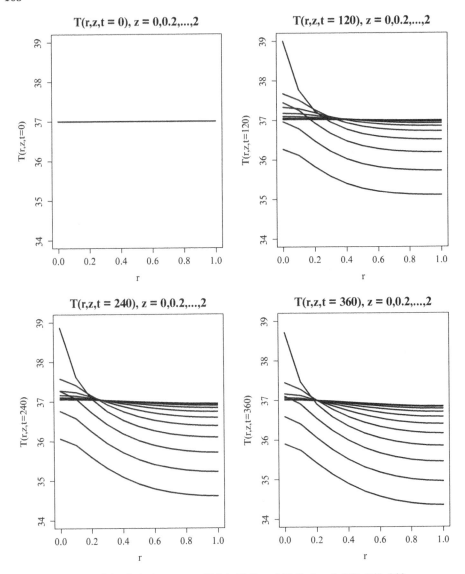

Figure 3.1   Solution to eqs. (3.1c), (3.2) and (3.3), $t = 0, 120, 240, 360$

The temperature rise at the skin surface $z = 0$ is of particular interest since that is where the temperature image is observed. To investigate this region in the thermal region, the output from the main program of Listing 3.1 is modified to

```
for(it in 1:nout){
  for(i in 1:nr){
    T[i,j]=out[it,(i-1)*nr+j+1];
    cat(sprintf("\n t = %3.0f    r = %4.2f    z = %4.2f
```

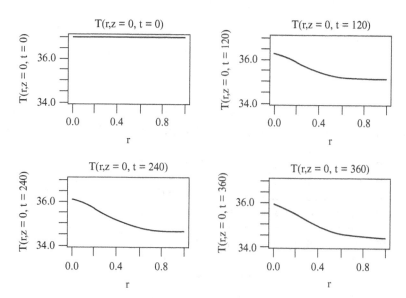

Figure 3.2 Solution to eqs. (3.1c), (3.2) and (3.3), $t = 0, 120, 240, 360, z = 0$

```
T = %5.2f",tout[it],r[i],z[1],T[i,1]));
    }
}
```

The skin surface temperature $T(r, z = 0, t)$ is now placed in arrays T, T1, T2, T3, T4 for subsequent plotting (note the subscript 1 in z[1], T[i,1] corresponding to $z = 0$). Only two nested for's are used (in t and r) since $z = 0$ and a for in z is not required. Also, the vertical scaling in the plotting is changed from ylim=c(34,39) to ylim=c(35,37) to accommodate the reduced temperature variation at $z = 0$.

Abbreviated numerical output is given in Table 3.2, and graphical output is given in Fig. 3.2.

We can note the following details about the output in Table 3.2.

- At $t = 360$, the maximum temperature, 35.90, is at $r = 0, z = 0$, as expected. This contrasts with the maximum temperature at the tumor, $T(r = 0, z = 0.6, t = 360) = 39.37$ (from Table 3.1).
- Temperatures fall below the IC temperature of $T(r, z = 0, t = 0) = 37.00$ °C due to the ambient temperature $T_a = 70$ °F = 21.1 °C.
- $T(r, z = 0, t)$ appears to reach a steady state in Fig. 3.2 as expected (from Fig. 3.1), but the steady-state temperatures $T(r, z = 0, t \to \infty)$ do not reach the ambient temperature, $T_a = 21.1$ °C, since a temperature drop at the skin surface is required to produce the steady-state heat transfer rate. This temperature drop is dependent on the heat transfer coefficient h in BC (3.3c). As the heat transfer coefficient increases, the temperature drop $(T_a - T(r, z = 0, t))$ decreases (this can be observed by executing the main program with an increased h).

```
nr = 11    nz = 11
nrow(out)
[1] 4
ncol(out)
[1] 122
ncall =     286
t =    0   r = 0.00   z = 0.00   T = 37.00
t =    0   r = 0.10   z = 0.00   T = 37.00
t =    0   r = 0.20   z = 0.00   T = 37.00
t =    0   r = 0.30   z = 0.00   T = 37.00
t =    0   r = 0.40   z = 0.00   T = 37.00
t =    0   r = 0.50   z = 0.00   T = 37.00
t =    0   r = 0.60   z = 0.00   T = 37.00
t =    0   r = 0.70   z = 0.00   T = 37.00
t =    0   r = 0.80   z = 0.00   T = 37.00
t =    0   r = 0.90   z = 0.00   T = 37.00
t =    0   r = 1.00   z = 0.00   T = 37.00

                  .          .
                  .          .
                  .          .
     Output for t = 120, 240 removed
                  .          .
                  .          .

                  .          .
t = 360    r = 0.00   z = 0.00   T = 35.90
t = 360    r = 0.10   z = 0.00   T = 35.73
t = 360    r = 0.20   z = 0.00   T = 35.42
t = 360    r = 0.30   z = 0.00   T = 35.13
t = 360    r = 0.40   z = 0.00   T = 34.89
t = 360    r = 0.50   z = 0.00   T = 34.71
t = 360    r = 0.60   z = 0.00   T = 34.57
t = 360    r = 0.70   z = 0.00   T = 34.47
t = 360    r = 0.80   z = 0.00   T = 34.40
t = 360    r = 0.90   z = 0.00   T = 34.36
t = 360    r = 1.00   z = 0.00   T = 34.35
```

Table 3.2: Abbreviated Output from Listings 3.1 and 3.2 for $z = 0$

(3.4) Summary and conclusions

The preceding discussion of the thermographic model of eqs. (3.1c), (3.2), and (3.3) indicates that an elevated skin surface temperature is possible due to heat release from a tumor. This elevated temperature could be observed with an infrared sensor to locate a tumor.

The model could be used to investigate how the skin temperature varies with the tumor location (the preceding results are for $z = 0.6$ cm) and thereby predict the effectiveness of locating a tumor at varying distances from the skin surface. The size of the tumor could be varied by using more than one grid point to define the tumor location.

Other interesting parameters of the model that could be investigated include: (1) Q_m, the tumor volumetric heat release rate; (2) k, the tissue heat conductivity; (3) k_e, the blood perfusion rate; and (4) h, the skin-surface-to-ambient-air heat transfer coefficient. In this way, the design and performance of a proposed thermographic tumor location system can be evaluated.

Finally, the accuracy of the numerical solutions is an important consideration. Since an analytical solution to eqs. (3.1c), (3.2), and (3.3) is not readily available, the solution accuracy can be accessed indirectly by:

- h-refinement in which the grid spacing in r and z is changed, e.g., by increasing the number of grid points, nr,nz, and observing the effect on the numerical solution. The name for this procedure comes from the use of the symbol h for the grid spacing in the numerical analysis literature. As a word of warning, as nr,nz are increased, the total number of MOL/ODEs increases (rapidly) as nr*nz.
- p-refinement in which the calculation of the spatial derivatives in eq. (3.1c) is performed with finite difference approximations (FDs) of varying order and the effect on the numerical solution is observed. The name for this procedure comes from the use of the symbol p for the order of the approximation in the numerical analysis literature. Changing the order is easily accomplished. For example, in thermo_1, dss044, which is based on fourth order FDs ($p = 4$), can be replaced with dss046, based on sixth order FDs ($p = 6$). The higher order approximations might have longer computer execution times since the number of calculations is higher, but this is usually a small effect, and experience has indicated that the use of higher order approximations is well worth the additional calculations.

h and p-refinement do not require an analytical solution (which is generally unavailable). This type of *error analysis* should be part of any numerical study to lend confidence to the accuracy (convergence) of the computed solutions. The preceding discussion pertains to the *truncation error* of the spatial approximations (the term truncation comes from the truncated Taylor series on which the FD approximations are based).

The same considerations of accuracy pertain to the integration in t (by ode). Generally, quality library ODE integrators produce solutions that satisfy the error tolerances (default 10^{-6} absolute and relative error tolerances of ode), or issue warning messages if the error tolerances are not met. Thus, the principal source of numerical error in MOL/PDE analysis is usually the spatial (boundary value) discretizations.

References

[1] Agnelli, J.P., A.A. Barrea, and C.V. Turner (2011), Tumor location and parameter estimation by thermography, *Mathematical and Computer Modelling*, **53**, 7-8, 1527–1534.

[2] Liu, E.H., G.M. Saidel, and H. Harasaki (2003), Model analysis of tissue responses to transient and chronic heating, *Annals of Biomedical Engineering*, **31**, 1007–1014.

[3] Schiesser, W.E. and G.W. Griffiths (2009), *A Compendium of Partial Differential Equation Models*, Cambridge University Press, Cambridge.

[4] Schiesser, W.E. (2014), *Differential Equation Analysis in Biomedical Science and Engineering: Partial Differential Equation Applications in R*, John Wiley & Sons, Inc., Hoboken, NJ.

4

BLOOD-TISSUE TRANSPORT

This chapter extends the discussion of first-order hyperbolic partial differential equations (PDEs) in Chapter 1 by introducing a 1D, 2-PDE model for a blood-tissue interface. Two cases of a *Riemann problem*[1] are considered: (1) a set of PDE parameters for which the numerical resolution of a discontinuity is easily accomplished with any of four approximations and (2) a set of PDE parameters for which the discontinuity remains pronounced and relatively difficult to resolve, which, therefore, provides a stringent test of the four approximations.

The intent of this chapter is to:

- Present a basic PDE model for blood-tissue transport, including the required initial conditions (ICs) and boundary conditions (BCs).
- Illustrate the coding of the model through a series of R routines, including the use of library routines for integration of the PDE derivatives in time and space.
- Present the computed model solution in numerical and graphical (plotted) formats.
- Compare the numerical and analytical solutions to give the exact error in the numerical solution.
- Evaluate several spatial derivative approximations through examination of the exact error in the numerical solution.
- Discuss the features of the numerical solution and the performance of the algorithms used to compute the solution.
- Consider extensions of the model.

[1]A Riemann problem is generally defined here as a first-order hyperbolic PDE system with discontinuous initial-boundary conditions.

Method of Lines PDE Analysis in Biomedical Science and Engineering, First Edition. William E. Schiesser.
© 2016 John Wiley & Sons, Inc. Published 2016 by John Wiley & Sons, Inc.
Companion website: www.wiley.com/go/Schiesser/PDE_Analysis

Concentration in u_1, u_2 possibly pertains to oxygen, nutrients for metabolism, and metabolites or toxins resulting from metabolism. We start with a statement of the model.

(4.1) 1D 2-PDE model

The 1D, 2-PDE blood-interface transport model is taken from [1,6],

$$\frac{\partial u_1}{\partial t} = -v\frac{\partial u_1}{\partial z} + k_1(u_2 - u_1) \tag{4.1a}$$

$$\frac{\partial u_2}{\partial t} = k_2(u_1 - u_2) \tag{4.1b}$$

where

u_1	blood concentration
u_2	tissue concentration
z	distance along the blood vessel
t	time
v	blood velocity
$k_1 k_2$	mass transfer coefficients

Eqs. (4.1) are derived in [6], pp 195–197.

Eq. (4.1a) is first order in t and z and therefore requires one IC and one BC.

$$u_1(z, t = 0) = f_1(z); \quad u_1(z = 0, t) = g_1(t) \tag{4.2a,b}$$

Eq. (4.1b) is first order in t and requires one IC.

$$u_2(z, t = 0) = f_2(z) \tag{4.2c}$$

Eqs. (4.1) and (4.2) are the 1D 2-PDE model to be studied by a MOL analysis. Since these equations are linear, an analytical solution is also available to evaluate the accuracy of the numerical solution.

The following analytical solution is derived in [6], p. 213.

$$u_{1a}(z, t) = u_a(z, t) + \frac{1}{k_2}\frac{\partial u_a(z, t)}{\partial t} \tag{4.3a}$$

where

$$u_a(z, t) = e^{-(k_1/v)z}k_2\int_0^t h(\lambda - z/v)e^{-k_2(\lambda - z/v)}I_o\left\{2\sqrt{\frac{k_1 k_2}{v}z(\lambda - z/v)}\right\}d\lambda \tag{4.3b}$$

and from *Leibniz's rule* for differentiating an integral (applied to eq. (4.3b))

$$\frac{\partial u_a(z, t)}{\partial t} = e^{-(k_1/v)z}k_2 h(t - z/v)e^{-k_2(t - z/v)}I_o\left\{2\sqrt{\frac{k_1 k_2}{v}z(t - z/v)}\right\} \tag{4.3c}$$

where I_o is the *modified Bessel function of the first kind of order zero* and $h(t - z/v)$ is the *Heaviside function* or unit step

$$h(t - z/v) = \begin{cases} 0, (t - z/v) < 0 \\ 1, (t - z/v) > 0 \end{cases} \tag{4.3d}$$

At $z = z_L$, eq. (4.3c) becomes an ODE in t

$$\frac{du_a(z = z_L, t)}{dt} = e^{-(k_1/v)z_L} k_2 h(t - z_L/v) e^{-k_2(t-z_L/v)} I_o \left\{ 2\sqrt{\frac{k_1 k_2}{v} z(t - z_L/v)} \right\} \tag{4.3e}$$

which is integrated numerically by adding the RHS function to the ODE routine pde_1 (discussed next). Thus, the analytical solution that is used to evaluate the numerical solution is itself partially numerical; that is, eq. (4.3e) is integrated numerically to facilitate the analysis. Eq. (4.3e) could also be integrated analytically, but that is relatively complicated and the numerical integration is straightforward.

(4.2) MOL routines

The MOL/ODE routine for ODE/PDE system of eqs. (4.1), (4.2b), and (4.3e) is considered first.

(4.2.1) MOL/ODE routine

The MOL/ODE routine is in Listing 4.1.

```
  pde_1=function(t,u,parms){
#
# Function pde_1 computes the t derivative
# vector of the u vector
#
# One vector to two PDEs
  u1=rep(0,n);u2=rep(0,n);
  for(i in 1:n){
    u1[i]=u[i];
    u2[i]=u[i+n];
  }
#
# Boundary condition
  u1[1]=1;
#
# First order spatial derivative
#
#   ifd = 1: Two-point upwind finite difference (2pu)
    if(ifd==1){ u1z=dss012(0,zL,n,u1,v); }
#
```

```
#    ifd = 2: Five-point center finite difference (5pc)
     if(ifd==2){ u1z=dss004(0,zL,n,u1); }
#
#    ifd = 3: Five-point biased upwind approximation (5pbu)
     if(ifd==3){ u1z=dss020(0,zL,n,u1,v); }
#
#    ifd = 4: van Leer flux limiter
     if(ifd==4){ u1z=vanl(0,zL,n,u1,v); }
#
# Temporal derivatives
     u1t=rep(0,n); u2t=rep(0,n);
#
#    u1t, u2t, u1at
     for(i in 1:n){
       if(i==1){
         u1t[i]=0;
       }else{
         u1t[i]=-v*u1z[i]+k1*(u2[i]-u1[i]);
       }
         u2t[i]=k2*(u1[i]-u2[i]);
     }
#
# Analytical ODE
  lam=t-zL/v;
  if(lam<0){
    uat=0;
  }else{
    arg=2*sqrt((k1*k2*zL/v)*lam)
    uat=exp1*k2*exp(-k2*lam)*bessel_Io(arg)
  }
#
# Two PDEs and one ODE to one derivative vector
  ut=rep(0,2*n);
  for(i in 1:n){
      ut[i]=u1t[i];
    ut[i+n]=u2t[i];
  }
  ut[2*n+1]=uat;
#
# Increment calls to pde_1
  ncall<<-ncall+1;
#
# Return derivative vector
  return(list(c(ut)));
}
```

Listing 4.1: ODE/PDE routine for eqs. (4.1), (4.2b), and (4.3e)

We can note the following details about Listing 4.1.

- The function is defined.

```
pde_1=function(t,u,parms){
#
# Function pde_1 computes the t derivative
# vector of the u vector
```

The input arguments are: (1) the current value of t, (2) a vector of dependent variables of length $2n + 1$ where $n = 41$ grid points in z set in the main program (discussed subsequently), and (3) a list of parameters available for use in pde_1 (unused in this case). n is available to pde_1 without any special designation, a feature of R.

- u is placed in two vectors, u1, u2, to facilitate the programming of eqs. (4.1). These vectors are first declared (preallocated) with the rep utility.

```
#
# One vector to two PDEs
  u1=rep(0,n);u2=rep(0,n);
  for(i in 1:n){
    u1[i]=u[i];
    u2[i]=u[i+n];
  }
```

- BC (4.2b) is applied as a constant $g_1(t) = 1$.

```
#
# Boundary condition
  u1[1]=1;
```

Note the use of subscript 1 corresponding to $z = 0$. This BC is equivalent to imposing the unit step of eq. (4.3d) with $z = 0$. In other words, the solution to be computed is the response to a unit step at $z = 0$ since the ICs (4.2a) and (4.2c) are *homogeneous* (zero) (as defined in the main program). BC (4.2b) is termed *Dirichlet* since the dependent variable $u_1(z = 0, t)$ is defined.

- The derivative $\dfrac{\partial u_1}{\partial z}$ in eq. (4.1a) is computed by one of four approximations as discussed in Chapter 1. ifd is set in the main program.

```
#
# First order spatial derivative
#
#    ifd = 1: Two-point upwind finite difference (2pu)
    if(ifd==1){ u1z=dss012(0,zL,n,u1,v); }
#
```

```
#    ifd = 2: Five-point center finite difference (5pc)
     if(ifd==2){ u1z=dss004(0,zL,n,u1); }
#
#    ifd = 3: Five-point biased upwind approximation (5pbu)
     if(ifd==3){ u1z=dss020(0,zL,n,u1,v); }
#
#    ifd = 4: van Leer flux limiter
     if(ifd==4){ u1z=vanl(0,zL,n,u1,v); }
```

- The derivatives $\dfrac{\partial u_1}{\partial t} = \text{u1t}$, $\dfrac{\partial u_2}{\partial t} = \text{u2t}$, in eqs. (4.1) are computed over the grid in z with a for. The programming is similar to eqs. (4.1), which is a principal advantage of the MOL.

```
#
# Temporal derivatives
    u1t=rep(0,n); u2t=rep(0,n);
#
#    u1t, u2t, u1at
     for(i in 1:n){
       if(i==1){
         u1t[i]=0;
       }else{
         u1t[i]=-v*u1z[i]+k1*(u2[i]-u1[i]);
       }
         u2t[i]=k2*(u1[i]-u2[i]);
     }
```

Since BC (4.2b) is a constant, its derivative in t (at $z = 0$) is set to zero so the ODE integrator will not change $u_1(z = 0, t) = 1$.

- ODE (4.3e) is programmed.

```
#
# Analytical ODE
    lam=t-zL/v;
    if(lam<0){
      uat=0;
    }else{
      arg=2*sqrt((k1*k2*zL/v)*lam)
      uat=exp1*k2*exp(-k2*lam)*bessel_Io(arg)
    }
```

This programming of eq. (4.3e) gives $\dfrac{du_a(z,t)}{dt} = \text{uat}$. $e^{-(k_1/v)z_L} = \text{exp1}$ is computed once in the main program since it is a constant. bessel_Io(arg) is a call to a function to compute I_o discussed subsequently.

Note the use of the *Lagrangian variable* $\lambda = (t - z_L/v) = $ lam (also used in eq. (4.3d)). The if expresses the discontinuity from $h(t - z_L/v)$ (in the RHS of eq. (4.3e)) that will be observed in the numerical solution.

- The two derivative vectors, u1t, u2t, and the ODE derivative uat are placed in a single derivative vector ut of length 2*n+1 = 2*41 + 1 = 83 for return to the ODE integrator ode called in the main program.

```
#
# Two PDEs and one ODE to one derivative vector
  ut=rep(0,2*n);
  for(i in 1:n){
      ut[i]=u1t[i];
    ut[i+n]=u2t[i];
  }
  ut[2*n+1]=uat;
```

- The counter for the calls to pde_1 is incremented and returned to the main program with <<-.

```
#
# Increment calls to pde_1
  ncall<<-ncall+1;
```

- The derivative vector ut is returned to the ODE integrator ode.

```
#
# Return derivative vector
  return(list(c(ut)));
}
```

c is the R utility for a vector. The derivative vector must be a list (required by the R ODE integrators). The final } concludes pde_1.

The main program is considered next.

(4.2.2) Main program

The main program that calls ode to compute the MOL solution and displays the numerical and analytical solutions follows.

```
#
# Delete previous workspaces
  rm(list=ls(all=TRUE))
#
#    1D, 2-PDE blood-tissue transport model
```

```
#
#   The PDE system is
#
#   u1_t = -v*u1_z + k1*(u2 - u1)                        (4.1a)
#
#   u2_t = k2*(u1 - u2)                                  (4.1b)
#
#   Initial conditions
#
#     u1(z,t=0) = f1(z)                                  (4.2a)
#
#     u2(z,t=0) = f2(z)                                  (4.2c)
#
#   Boundary condition
#
#     u1(z=0,t) = g1(t)                                  (4.2b)
#
#   The method of lines (MOL) solution for eqs. (4.1),
#   (4.2) is coded below.  Specifically, the spatial
#   derivative in the fluid balance, u1_z in eq. (4.1a),
#   is replaced by one of four approximations as selected
#   by the variable ifd.
#
# Access ODE integrator
  library("deSolve");
#
# Access files
  setwd("g:/comp3/chap4");
  source("pde_1.R") ;source("bessel_Io.R")  ;
  source("dss004.R");source("dss012.R")       ;
  source("dss020.R");source("van1.R")         ;
#
# Step through cases
  for(ncase in 1:4){
#
# Model parameters
    u10=0;   u20=0;
     k1=1;    k2=1;
# k1=0.1; k2=0.1;
      v=1;    zL=5;
     n=41;
    exp1=exp(-k1/v*zL);
#
# Select an approximation for the convective derivative u1z
#
#   ifd = 1: Two-point upwind approximation
```

```
#
#   ifd = 2: Centered approximation
#
#   ifd = 3: Five-point, biased upwind approximation
#
#   ifd = 4: van Leer flux limiter
#
  ifd=ncase;
#
# Level of output
#
#   Detailed output   - ip = 1
#
#   Brief (IC) output - ip = 2
#
  ip=1;
#
# Initial condition
  u0=rep(0,2*n+1);
  for(i in 1:n){
      u0[i]=u10;
    u0[i+n]=u20;
    }
  u0[2*n+1]=u10;
#
# Grid in t
  t0=0;tf=20;nout=41;
  tout=seq(from=t0,to=tf,by=(tf-t0)/(nout-1));
  ncall=0;
#
# ODE integration
  out=ode(func=pde_1,times=tout,y=u0);
#
# Store solution
  u1=matrix(0,nrow=nout,ncol=n);
  u2=matrix(0,nrow=nout,ncol=n);
  uat=rep(0,nout);ua=rep(0,nout);
  u1a=rep(0,nout);t=rep(0,nout);
  for(it in 1:nout){
  for(iz in 1:n){
    u1[it,iz]=out[it,iz+1];
    u2[it,iz]=out[it,iz+1+n];
  }
    ua[it]=out[it,2*n+2];
     t[it]=out[it,1];
  }
```

```
    cat(sprintf("\n\n ifd = %2d   ncase = %2d   ncall = %2d",
                ifd,ncase,ncall));
#
# Display numerical solution
  if(ip==1){
    cat(sprintf(
      "\n\n      t      u1(z=zL,t)  u1a(z=zL,t)      diff\n"));
    for(it in 1:nout){
      lam=t[it]-zL/v;
      if(lam<0){
        uat=0;
      }else{
        arg=2*sqrt((k1*k2*zL/v)*lam)
        uat=exp1*k2*exp(-k2*lam)*bessel_Io(arg)
      }
      u1a[it]=ua[it]+uat/k2;
      diff=u1[it,n]-u1a[it];
      cat(sprintf(
        "%7.2f%12.4f%12.4f%14.6f\n",
        t[it],u1[it,n],u1a[it],diff));
    }
  }
#
# Plot analytical, numerical solutions
  par(mfrow=c(1,1))
  plot(t,u1[,n],xlab="t",ylab="u1(z=zL,t)",
       main="u1(z=zL,t), solid - anal, o - num",
       col="black",lwd=2,pch="o");
  lines(t,u1[,n],lty=2,lwd=2,type="l");
  lines(t,u1a,lty=1,lwd=2,type="l");
#
# Next case
  }
```

Listing 4.2: Main program for eqs. (4.1) to (4.3)

We can note the following details about this main program in Listing 4.2.

- Previous workspaces are removed. The documentation comments for the problem statement are not repeated here to conserve space.

```
#
# Delete previous workspaces
  rm(list=ls(all=TRUE))
```

- The ODE library desolve is accessed to provide the integrator ode. The files used in the subsequent programming are accessed with the source utility. The setwd (set working directory) requires editing for the local computer (note the use of / rather than the usual \).

```
#
# Access ODE integrator
  library("deSolve");
#
# Access files
  setwd("g:/comp3/chap4");
  source("pde_1.R") ;source("bessel_Io.R")  ;
  source("dss004.R");source("dss012.R")      ;
  source("dss020.R");source("van1.R")        ;
```

- Four cases are executed corresponding to four approximations of the derivative $\frac{\partial u_1}{\partial z}$ in eq. (4.1a).

```
#
# Step through cases
  for(ncase in 1:4){
```

- The model parameters are defined numerically.

```
#
# Model parameters
    u10=0;   u20=0;
     k1=1;    k2=1;
# k1=0.1; k2=0.1;
      v=1;    zL=5;
      n=41;
    exp1=exp(-k1/v*zL);
```

In particular,

- Two cases are programmed through changes in the mass transfer coefficients k_1, k_2. For the first case, $k_1 = 1, k_2 = 1$, mass transfer from the blood to the tissue is large enough that the moving front in u_1 is smoothed so that the four FD approximations for the derivative $\frac{\partial u_1}{\partial z}$ in eq. (4.1a) (programmed in Listing 4.1), all give essentially the same solution that is in agreement with the analytical solution of eq. (4.3a).
- For the second case, $k_1 = 0.1, k_2 = 0.1$ (from removing #), the u_1 moving front is sharpened so that the four approximations give differing results, and only the van Leer flux limiter (ifd = 4 in Listing 4.1) gives a solution with minimum numerical diffusion and oscillation, which is in good agreement with the analytical solution of eq. (4.3a). These results are discussed subsequently.

- The constant exp1=exp(-k1/v*zL) is computed for subsequent use, including in pde_1 in Listing 4.1, which receives the value without any special designation or programming (a feature of R).
- The time for the front in u_1 to move from $z = 0$ (where the unit step is applied) to $z = z_L$ is approximately $z_L/v = 5/1 = 5$, which is observed in the numerical solution, particularly for the second case.

- One of the four approximations for $\dfrac{\partial u_1}{\partial z}$ in eq. (4.1a) is selected with the index ncase of the preceding for.

```
#
# Select an approximation for the convective derivative u1z
#
#    ifd = 1: Two-point upwind approximation
#
#    ifd = 2: Centered approximation
#
#    ifd = 3: Five-point, biased upwind approximation
#
#    ifd = 4: van Leer flux limiter
#
    ifd=ncase;
```

- The level of output is selected. In this case, ip=1 gives the numerical output of the solution, as well as the graphical (plotted) output (discussed subsequently).

```
#
# Level of output
#
#    Detailed output   - ip = 1
#
#    Brief (IC) output - ip = 2
#
    ip=1;
```

- The IC vector for the $2n + 1 = 83$ ODEs is defined. In this case, the ICs are homogeneous since u10 = u20 = 0 set previously. The ICs are for eqs. (4.2a), (4.2c), and (4.3e).

```
#
# Initial condition
    u0=rep(0,2*n+1);
    for(i in 1:n){
        u0[i]=u10;
        u0[i+n]=u20;
    }
    u0[2*n+1]=u10;
```

- The interval in t, $0 \le t \le 20$, is defined with 41 output points (including $t = 0$) so the output interval is $(20 - 0)/(41 - 1) = 0.5$. The seq utility is used to define the sequence $t = 0, 0.5,...,20$ in tout.

```
#
# Grid in t
  t0=0;tf=20;nout=41;
  tout=seq(from=t0,to=tf,by=(tf-t0)/(nout-1));
  ncall=0;
```

The counter for the calls to pde_1 is also initialized.
- The 83 ODEs are integrated by ode.

```
#
# ODE integration
  out=ode(func=pde_1,times=tout,y=u0);
```

The arguments to ode are: (1) the MOL/ODE routine pde_1 of Listing 4.1, (2) the vector of output values of t, tout, and (3) the IC vector u0. func,times,y are reserved names. The numerical solution is returned in matrix out. The length of u0 (83) informs ode of the number of ODEs to be integrated.
- The numerical solution in out is placed in matrices and vectors for subsequent display numerically and graphically.

```
#
# Store solution
  u1=matrix(0,nrow=nout,ncol=n);
  u2=matrix(0,nrow=nout,ncol=n);
  uat=rep(0,nout);ua=rep(0,nout);
  u1a=rep(0,nout);t=rep(0,nout);
  for(it in 1:nout){
  for(iz in 1:n){
    u1[it,iz]=out[it,iz+1];
    u2[it,iz]=out[it,iz+1+n];
  }
    ua[it]=out[it,2*n+2];
     t[it]=out[it,1];
  }
```

In particular,
- The arrays for u_1, u_2 are 2D for z, t. The offset 1 in iz+1,iz+1+n is used since the first position in out is for t, that is, t[it]=out[it,1].
- The vectors for the analytical solution of eq. (4.3a), uat,ua,u1a, are 1D for t. The solution of eq. (4.3e) is placed in the vector ua, that is, ua[it]=out[it,2*n+2]. The use of these vectors is explained subsequently.

- The dimensions of out are therefore out (21,84). Note in particular the second dimension 2*n+2 = 2*41+2 = 84.

- The counter for the calls to pde_1 of Listing 4.1 is displayed as a measure of the computational effort required to compute the numerical solution.

```
#
  cat(sprintf("\n\n ifd = %2d   ncase = %2d   ncall = %2d",
              ifd,ncase,ncall));
```

- The numerical solution is displayed for ip=1.

```
#
# Display numerical solution
  if(ip==1){
    cat(sprintf(
      "\n\n     t      u1(z=zL,t)  u1a(z=zL,t)     diff\n"));
    for(it in 1:nout){
      lam=t[it]-zL/v;
      if(lam<0){
        uat=0;
      }else{
        arg=2*sqrt((k1*k2*zL/v)*lam)
        uat=exp1*k2*exp(-k2*lam)*bessel_Io(arg)
      }
      u1a[it]=ua[it]+uat/k2;
      diff=u1[it,n]-u1a[it];
      cat(sprintf(
        "%7.2f%12.4f%12.4f%14.6f\n",
        t[it],u1[it,n],u1a[it],diff));
    }
  }
```

In particular,

- A heading is displayed for the numerical solution u_1 at $z = z_L$ (from eq. (4.1a)), the analytical solution u_{1a} from eq. (4.3a), and the difference of the two solutions (the exact error in the numerical solution).

- du_a/dt from eq. (4.3e) is computed since it is part of the analytical solution of eq. (4.3a). This code essentially duplicates the code for du_a/dt in pde_1 of Listing 4.1. The exception is using the vector t[it] rather than the scalar t in pde_1.

```
    for(it in 1:nout){
      lam=t[it]-zL/v;
      if(lam<0){
        uat=0;
      }else{
        arg=2*sqrt((k1*k2*zL/v)*lam)
        uat=exp1*k2*exp(-k2*lam)*bessel_Io(arg)
      }
```

– The analytical solution of eq. (4.3a) is computed, then the exact error in the
numerical solution is computed as diff.

```
        u1a[it]=ua[it]+uat/k2;
        diff=u1[it,n]-u1a[it];
        cat(sprintf(
          "%7.2f%12.4f%12.4f%14.6f\n",
          t[it],u1[it,n],u1a[it],diff));
    }
  }
```

The analytical and numerical solutions, and their difference, are displayed. The
two }'s terminate the for in it (in t) and the if for ip=1.
• The analytical and numerical solutions are plotted as a solid line and as points,
respectively, against t (see Fig. 4.1 and subsequent plots).

```
#
# Plot analytical, numerical solutions
  par(mfrow=c(1,1))
  plot(t,u1[,n],xlab="t",ylab="u1(z=zL,t)",
      main="u1(z=zL,t), solid - anal, o - num",
      col="black",lwd=2,pch="o");
  lines(t,u1[,n],lty=2,lwd=2,type="l");
  lines(t,u1a,lty=1,lwd=2,type="l");
```

Vectorization is used as u1[,n] for all values of t at $z = z_L$. The points for the
numerical solution are specified with pch="o". A connection of the points is

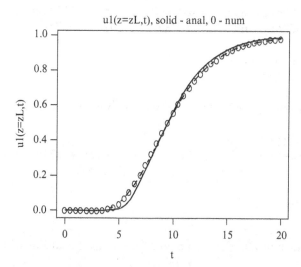

Figure 4.1   Analytical and numerical solutions of eqs. (4.1), ifd=1, $k_1 = k_2 = 1$

specified with the first call to `lines`, which produced a dashed line with `lty=2`. The analytical solution of eq. (4.3a) is plotted as a solid line with a second call to `lines` and `lty=1`.

- The `for` in `ncase`, `for(ncase in 1:4){`, is concluded.

```
#
# Next case
  }
```

This concludes the discussion of the main program of Listing 4.2. The routine for the modified Bessel function in the analytical solution of eqs. (4.3a) and (4.3e) concludes the programming.

(4.2.3) Bessel function routine

The routine for I_o is listed next.

```
bessel_Io=function(x){
#
# Function bessel_Io is a translation from Fortran
# into R of the original numerical recipes function
# BESSIO.
#
# Constants in the approximation of the Bessel
# function Io
  p1=1              ; p2=3.5156229    ; p3=3.0899424    ;
  p4=1.2067492    ; p5=0.2659732    ; p6=0.360768e-1  ;
  p7=0.45813e-2   ;
  q1=0.39894228   ; q2=0.1328592e-1 ; q3=0.225319e-2  ;
  q4=-0.157565e-2; q5=0.916281e-2   ; q6=-0.2057706e-1;
  q7=0.2635537e-1; q8=-0.1647633e-1; q9=0.392377e-2   ;
#
# Calculation of bessel_Io
  if(abs(x)<3.75){
    y=(x/3.75)^{2};
    bessel_Io=p1+y*(p2+y*(p3+y*(p4+y*(p5+y*(p6+y*p7)))));
  }else{
    ax=abs(x);y=3.75/ax;
    bessel_Io=(exp(ax)/sqrt(ax))*(q1+y*(q2+y*(q3+y*(q4+
              y*(q5+y*(q6+y*(q7+y*(q8+y*q9))))))));
  }
  return(c(bessel_Io));
  }
```

Listing 4.3: Function `bessel_Io` for the calculation of I_o

The details of this routine will not be discussed [4]. Briefly, I_o is defined by an infinite series that is approximated by a combination of exponential and polynomial functions. The computed Bessel function is returned as a numerical value with `return(c(bessel_Io))` (a `list` is not required as in `pde_1` of Listing 4.1).

(4.3) Model output

The output from the preceding routines is discussed next. For the first case ($k_1 = k_2 = 1$), the numerical output is
We can note the following details of this output.

- With the exception of the case `ifd = 1`, the differences between the numerical and analytical solutions are generally small. The maximum difference in this output is for

```
ifd =  1    ncase =  1    ncall = 769
     t     u1(z=zL,t)  u1a(z=zL,t)      diff
  6.00       0.1065      0.0656     0.040907
```

which reflects the numerical diffusion introduced with the 2pu approximation (as indicated in Figs. 4.1 and 4.5).

- `ifd=1` has the smallest value of `ncall` ($= 769$), (but again, the poorest accuracy) while `ifd=4` has the largest value (`ncall = 13058`). Since the accuracy of the van Leer flux limiter (`ifd=4`) is somewhat less than for the 5pc FD `ifd=2` or the 5pbu FD `ifd=3`,[2] which required markedly fewer calls to `pde_1` (836 and 1072, respectively), then the question of why the van Leer limiter would be used arises. This question is answered next with the second case ($k_1 = k_2 = 0.1$), which demonstrates conclusively the best performance of the van Leer limiter.
- As an incidental note, several flux limiters have been reported ([3], pp. 37–43) and possibly some would perform better than the van Leer limiter used here. This could be easily investigated by replacing the statements in the van Leer routine, `van1`, with corresponding statements for the other limiters.

The graphical output for $k_1 = k_2 = 1$ is given in Figs. 4.1 to 4.4.

We can observe in Figs. 4.1 to 4.4 that 2pu (`ifd=1`) gives the poorest numerical solution due to numerical diffusion. The solutions for 5pc, 5pbu, and van Leer are accurate and indistinguishable.

For the second case, $k_1 = k_2 = 0.1$, the numerical solution has markedly different characteristics. The corresponding graphical output is shown in Figs. 4.5 to 4.8.

We can note the following features of the numerical output in Table 4.2 and Figs. 4.5 to 4.8.

[2] This follows from visual inspection of the output in Table 4.1, particularly the values of `diff`.

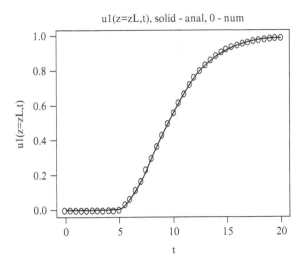

Figure 4.2   Analytical and numerical solutions of eqs. (4.1), `ifd=2`, $k_1 = k_2 = 1$

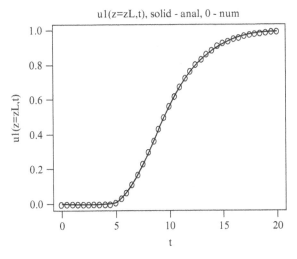

Figure 4.3   Analytical and numerical solutions of eqs. (4.1), `ifd=3`, $k_1 = k_2 = 1$

- The effect of the discontinuity $h(t - z_l/v)$ in eq. (4.3e) is clear. The programming of this discontinuity in the MOL/PDE routine of Listing 4.1 and main program of Listing 4.2 was straightforward (the switch when $t - z_L/v$ changed sign).
- For `ifd=1,2,3`, substantial differences between the analytical solution (eq. (4.3a)) and the numerical solution are apparent (see Figs. 4.5 to 4.7). Conversely, there is much closer agreement between the two solutions for `ifd=4` (see Fig. 4.8) indicating that the van Leer flux limiter has produced a relatively good solution.
  - For `ifd=1`, substantial numerical diffusion is apparent.

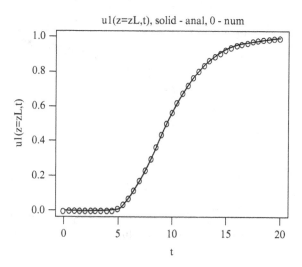

ul(z=zL,t), solid - anal, 0 - num

Figure 4.4 Analytical and numerical solutions of eqs. (4.1), `ifd=4`, $k_1 = k_2 = 1$

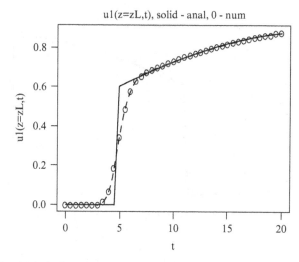

ul(z=zL,t), solid - anal, 0 - num

Figure 4.5 Analytical and numerical solutions of eqs. (4.1), `ifd=1`, $k_1 = k_2 = 0.1$

- For `ifd=2`, substantial numerical oscillation is apparent.
- For `ifd=3`, numerical oscillation still occurs at the left side of the moving front. This smaller oscillation might still be unacceptable in modeling physical systems that cannot oscillate or for which negative values are physically impossible such as for the concentration u_1 (as in Fig. 4.7).
- The largest errors in all four cases (`ifd=1,2,3,4`) occur when the moving front reaches $z = z_L$ (at $t = 5$) as might be expected since $t - z_L/v = 5 - 5/1 = 0$ (when the switch in the programming of the analytical solution occurs).

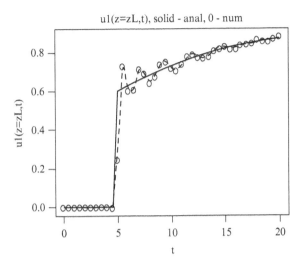

Figure 4.6   Analytical and numerical solutions of eqs. (4.1), `ifd=2`, $k_1 = k_2 = 0.1$

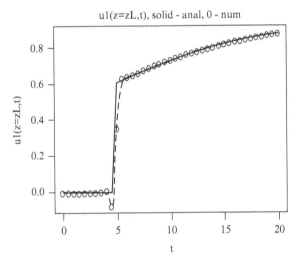

Figure 4.7   Analytical and numerical solutions of eqs. (4.1), `ifd=3`, $k_1 = k_2 = 0.1$

- `ifd=1` has the smallest value of `ncall` ($= 700$) (but again, with substantial numerical diffusion) while `ifd=4` has the largest value (`ncall = 2176`). Also, somewhat unexpectedly, for $k_1 = k_2 = 0.1$, the van Leer limiter required `ncall = 2176` calls to `pde_1`, which is substantially fewer than for $k_1 = k_2 = 1$ with `ncall = 13058` (from Table 4.1). This result may possibly follow from the observation that for the first case, the solution has a significant variation throughout $0 \leq z \leq z_L$ due to smoothing (see Fig. 4.4) while for the second case, the solution has a large variation only near the moving front, i.e., where $t - z_L/v = 0$ (Fig. 4.8). But some

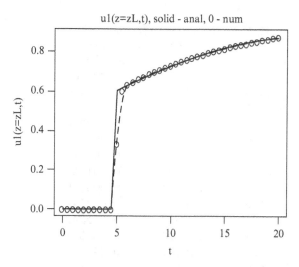

Figure 4.8 Analytical and numerical solutions of eqs. (4.1), `ifd=4`, $k_1 = k_2 = 0.1$

further analysis of the van Leer limiter is required to explain this result, and also, other limiters could be used for further study.

To conclude, the character of the solution to eqs. (4.1) is changed substantially just with the parameter variation $k_1 = k_2 = 1$ to $k_1 = k_2 = 0.1$ (everything else remains unchanged). Also, the performance of the approximations for $\dfrac{\partial u_1}{\partial z}$ in eq. (4.1a) changes substantially with the change in k_1, k_2.

(4.4) Model extensions

Eqs. (4.1) and (4.2) are linear, and therefore, an analytical solution, eq. (4.3a), is available to evaluate the error of the numerical solution. Nonlinear extensions of eqs. (4.1) have been proposed to describe a broader class of problems in blood-tissue transport, but for which analytical solutions are precluded (because of the nonlinearity and complexity of the model). For example, the following 1D, 2-PDE system has been studied [7]

$$\frac{\partial u_1}{\partial t} = -v\frac{\partial u_1}{\partial z} + f_1(u_1, u_2) \tag{4.4a}$$

$$\frac{\partial u_2}{\partial t} = f_2(u_1, u_2) \tag{4.4b}$$

Eqs. (4.4) have generalized transport functions f_1, f_2 that are nonlinear and involve additional variables.[3] For example, f_1 is

$$f_1(u_1, u_2) = -c_1 p - c_2 u_2 \tag{4.4c}$$

---

[3] f_2 was not specified in [7] and the computer code was not provided, so the reported numerical solution could not be verified. Also, the reported MOL solution has excessive numerical diffusion, possibly because of the use of 2pu FDs, but the computational details and code were not provided in [7].

```
ifd =   1    ncase =   1    ncall = 769
    t      u1(z=zL,t)  ua1(z=zL,t)      diff
  0.00       0.0000       0.0000      0.000000
  0.50       0.0000       0.0000      0.000000
  1.00       0.0000       0.0000      0.000000
  1.50       0.0000       0.0000      0.000000
  2.00       0.0000       0.0000      0.000000
  2.50       0.0000       0.0000      0.000006
  3.00       0.0002       0.0000      0.000153
  3.50       0.0014       0.0000      0.001351
  4.00       0.0061       0.0000      0.006105
  4.50       0.0176       0.0000      0.017649
  5.00       0.0381       0.0067      0.031342
  5.50       0.0679       0.0299      0.037943
  6.00       0.1065       0.0656      0.040907
  6.50       0.1529       0.1125      0.040363
  7.00       0.2054       0.1686      0.036866
  7.50       0.2625       0.2313      0.031181
  8.00       0.3223       0.2982      0.024119
  8.50       0.3832       0.3668      0.016432
  9.00       0.4438       0.4351      0.008758
  9.50       0.5028       0.5012      0.001587
 10.00       0.5592       0.5639     -0.004746
 10.50       0.6122       0.6222     -0.010041
 11.00       0.6613       0.6755     -0.014217
 11.50       0.7063       0.7236     -0.017282
 12.00       0.7470       0.7663     -0.019312
 12.50       0.7834       0.8038     -0.020425
 13.00       0.8157       0.8365     -0.020760
 13.50       0.8440       0.8645     -0.020467
 14.00       0.8687       0.8884     -0.019688
 14.50       0.8901       0.9086     -0.018558
 15.00       0.9084       0.9256     -0.017194
 15.50       0.9240       0.9397     -0.015692
 16.00       0.9373       0.9514     -0.014135
 16.50       0.9484       0.9610     -0.012584
 17.00       0.9578       0.9689     -0.011085
 17.50       0.9656       0.9752     -0.009672
 18.00       0.9720       0.9804     -0.008365
 18.50       0.9773       0.9845     -0.007178
 19.00       0.9817       0.9878     -0.006113
 19.50       0.9853       0.9905     -0.005171
 20.00       0.9882       0.9926     -0.004346
```

Table 4.1: Numerical Output from the Main Program of Listing 4.2, $k_1 = k_2 = 1$

```
ifd =  2    ncase =  2    ncall = 835
     t     u1(z=zL,t)  ua1(z=zL,t)      diff
   0.00     0.0000       0.0000      0.000000
   0.50    -0.0000       0.0000     -0.000000
   1.00     0.0000       0.0000      0.000000
   1.50     0.0000       0.0000      0.000000
   2.00     0.0000       0.0000      0.000000
   2.50    -0.0000       0.0000     -0.000000
   3.00     0.0000       0.0000      0.000000
   3.50    -0.0000       0.0000     -0.000000
   4.00     0.0000       0.0000      0.000012
   4.50    -0.0004       0.0000     -0.000359
   5.00     0.0047       0.0067     -0.002015
   5.50     0.0306       0.0300      0.000665
   6.00     0.0653       0.0656     -0.000338
   6.50     0.1126       0.1125      0.000026
   7.00     0.1687       0.1686      0.000102
   7.50     0.2313       0.2313     -0.000033
   8.00     0.2982       0.2982     -0.000041
   8.50     0.3668       0.3668      0.000007
   9.00     0.4351       0.4351      0.000011
   9.50     0.5012       0.5012     -0.000008
  10.00     0.5639       0.5639     -0.000004
  10.50     0.6222       0.6222      0.000006
  11.00     0.6755       0.6755      0.000005
  11.50     0.7236       0.7236      0.000003
  12.00     0.7663       0.7663      0.000008
  12.50     0.8038       0.8038      0.000009
  13.00     0.8365       0.8364      0.000007
  13.50     0.8645       0.8645      0.000009
  14.00     0.8884       0.8884      0.000010
  14.50     0.9087       0.9086      0.000009
  15.00     0.9256       0.9256      0.000009
  15.50     0.9397       0.9397      0.000008
  16.00     0.9514       0.9514      0.000009
  16.50     0.9610       0.9610      0.000007
  17.00     0.9689       0.9689      0.000006
  17.50     0.9752       0.9752      0.000006
  18.00     0.9804       0.9804      0.000005
  18.50     0.9845       0.9845      0.000004
  19.00     0.9878       0.9878      0.000003
  19.50     0.9905       0.9905      0.000003
  20.00     0.9926       0.9926      0.000003
```

Table 4.1: (*Continued*)

```
ifd =  3   ncase =  3   ncall = 1072
   t     u1(z=zL,t)  ua1(z=zL,t)      diff
 0.00      0.0000      0.0000      0.000000
 0.50      0.0000      0.0000      0.000000
 1.00      0.0000      0.0000      0.000000
 1.50      0.0000      0.0000      0.000000
 2.00     -0.0000      0.0000     -0.000001
 2.50     -0.0000      0.0000     -0.000007
 3.00     -0.0000      0.0000     -0.000011
 3.50      0.0002      0.0000      0.000246
 4.00      0.0001      0.0000      0.000052
 4.50     -0.0010      0.0000     -0.000954
 5.00      0.0055      0.0067     -0.001283
 5.50      0.0299      0.0299     -0.000004
 6.00      0.0656      0.0656     -0.000000
 6.50      0.1125      0.1125     -0.000003
 7.00      0.1686      0.1686      0.000001
 7.50      0.2313      0.2313      0.000007
 8.00      0.2982      0.2982      0.000009
 8.50      0.3668      0.3668      0.000010
 9.00      0.4351      0.4351      0.000010
 9.50      0.5012      0.5012      0.000009
10.00      0.5639      0.5639      0.000007
10.50      0.6222      0.6222      0.000005
11.00      0.6756      0.6755      0.000003
11.50      0.7236      0.7236      0.000002
12.00      0.7663      0.7663      0.000001
12.50      0.8038      0.8038      0.000000
13.00      0.8364      0.8364      0.000000
13.50      0.8645      0.8645     -0.000000
14.00      0.8884      0.8884      0.000000
14.50      0.9086      0.9086     -0.000000
15.00      0.9256      0.9256     -0.000000
15.50      0.9397      0.9397      0.000000
16.00      0.9514      0.9514     -0.000000
16.50      0.9610      0.9610     -0.000000
17.00      0.9689      0.9689     -0.000000
17.50      0.9752      0.9752     -0.000000
18.00      0.9804      0.9804     -0.000001
18.50      0.9845      0.9845      0.000001
19.00      0.9878      0.9878     -0.000001
19.50      0.9905      0.9905     -0.000000
20.00      0.9926      0.9926     -0.000000
```

Table 4.1: (*Continued*)

```
ifd =   4    ncase =   4    ncall = 13058
     t       u1(z=zL,t)   ua1(z=zL,t)      diff
   0.00        0.0000        0.0000      0.000000
   0.50       -0.0000        0.0000     -0.000000
   1.00       -0.0000        0.0000     -0.000000
   1.50       -0.0000        0.0000     -0.000000
   2.00       -0.0000        0.0000     -0.000000
   2.50       -0.0000        0.0000     -0.000000
   3.00       -0.0000        0.0000     -0.000000
   3.50       -0.0000        0.0000     -0.000000
   4.00       -0.0000        0.0000     -0.000000
   4.50       -0.0000        0.0000     -0.000000
   5.00        0.0073        0.0067      0.000594
   5.50        0.0324        0.0300      0.002431
   6.00        0.0686        0.0656      0.003006
   6.50        0.1157        0.1125      0.003221
   7.00        0.1717        0.1686      0.003169
   7.50        0.2342        0.2313      0.002898
   8.00        0.3007        0.2982      0.002469
   8.50        0.3688        0.3668      0.001944
   9.00        0.4365        0.4351      0.001380
   9.50        0.5020        0.5012      0.000824
  10.00        0.5642        0.5639      0.000311
  10.50        0.6221        0.6222     -0.000136
  11.00        0.6750        0.6755     -0.000507
  11.50        0.7228        0.7236     -0.000794
  12.00        0.7653        0.7663     -0.001002
  12.50        0.8027        0.8038     -0.001137
  13.00        0.8352        0.8364     -0.001209
  13.50        0.8633        0.8645     -0.001229
  14.00        0.8872        0.8884     -0.001208
  14.50        0.9075        0.9086     -0.001155
  15.00        0.9245        0.9256     -0.001079
  15.50        0.9387        0.9397     -0.000991
  16.00        0.9505        0.9514     -0.000895
  16.50        0.9602        0.9610     -0.000797
  17.00        0.9681        0.9688     -0.000699
  17.50        0.9746        0.9752     -0.000607
  18.00        0.9798        0.9804     -0.000522
  18.50        0.9841        0.9845     -0.000444
  19.00        0.9874        0.9878     -0.000373
  19.50        0.9901        0.9904     -0.000311
  20.00        0.9923        0.9925     -0.000257
```

Table 4.1: (*Continued*)

```
ifd =   1    ncase =   1    ncall = 700
     t     u1(z=zL,t)   ua1(z=zL,t)     diff
   0.00       0.0000       0.0000      0.000000
   0.50       0.0000       0.0000      0.000000
   1.00       0.0000       0.0000      0.000000
   1.50       0.0000       0.0000      0.000000
   2.00       0.0000       0.0000      0.000000
   2.50       0.0000       0.0000      0.000042
   3.00       0.0013       0.0000      0.001311
   3.50       0.0138       0.0000      0.013769
   4.00       0.0668       0.0000      0.066784
   4.50       0.1852       0.0000      0.185212
   5.00       0.3442       0.6065     -0.262290
   5.50       0.4867       0.6214     -0.134685
   6.00       0.5803       0.6357     -0.055492
   6.50       0.6306       0.6495     -0.018981
   7.00       0.6569       0.6628     -0.005916
   7.50       0.6735       0.6756     -0.002177
   8.00       0.6867       0.6880     -0.001318
   8.50       0.6987       0.6998     -0.001183
   9.00       0.7101       0.7113     -0.001195
   9.50       0.7211       0.7223     -0.001231
  10.00       0.7316       0.7329     -0.001264
  10.50       0.7418       0.7431     -0.001295
  11.00       0.7516       0.7529     -0.001321
  11.50       0.7610       0.7624     -0.001343
  12.00       0.7701       0.7714     -0.001362
  12.50       0.7788       0.7802     -0.001377
  13.00       0.7872       0.7886     -0.001389
  13.50       0.7954       0.7968     -0.001399
  14.00       0.8032       0.8046     -0.001405
  14.50       0.8107       0.8121     -0.001409
  15.00       0.8179       0.8193     -0.001410
  15.50       0.8249       0.8263     -0.001409
  16.00       0.8316       0.8330     -0.001407
  16.50       0.8380       0.8394     -0.001402
  17.00       0.8442       0.8456     -0.001395
  17.50       0.8502       0.8516     -0.001387
  18.00       0.8559       0.8573     -0.001378
  18.50       0.8615       0.8628     -0.001367
  19.00       0.8668       0.8681     -0.001355
  19.50       0.8719       0.8733     -0.001341
  20.00       0.8768       0.8782     -0.001327
```

Table 4.2: Numerical Output from the Main Program of Listing 4.2, $k_1 = k_2 = 0.1$

```
ifd =  2    ncase =  2    ncall = 1492
      t     u1(z=zL,t)   ua1(z=zL,t)      diff
    0.00      0.0000       0.0000       0.000000
    0.50     -0.0000       0.0000      -0.000000
    1.00      0.0000       0.0000       0.000000
    1.50      0.0000       0.0000       0.000000
    2.00      0.0000       0.0000       0.000000
    2.50     -0.0000       0.0000      -0.000000
    3.00      0.0000       0.0000       0.000000
    3.50     -0.0000       0.0000      -0.000003
    4.00      0.0002       0.0000       0.000194
    4.50     -0.0097       0.0000      -0.009708
    5.00      0.2449       0.6065      -0.361651
    5.50      0.7300       0.6214       0.108594
    6.00      0.6005       0.6357      -0.035290
    6.50      0.6071       0.6496      -0.042408
    7.00      0.7141       0.6628       0.051288
    7.50      0.6954       0.6756       0.019787
    8.00      0.6459       0.6880      -0.042054
    8.50      0.6772       0.6998      -0.022688
    9.00      0.7399       0.7113       0.028631
    9.50      0.7526       0.7223       0.030322
   10.00      0.7232       0.7329      -0.009633
   10.50      0.7124       0.7431      -0.030671
   11.00      0.7426       0.7529      -0.010271
   11.50      0.7812       0.7624       0.018844
   12.00      0.7926       0.7715       0.021120
   12.50      0.7795       0.7802      -0.000674
   13.00      0.7726       0.7886      -0.016027
   13.50      0.7821       0.7968      -0.014702
   14.00      0.8110       0.8046       0.006468
   14.50      0.8214       0.8121       0.009320
   15.00      0.8294       0.8193       0.010104
   15.50      0.8222       0.8263      -0.004053
   16.00      0.8215       0.8330      -0.011425
   16.50      0.8404       0.8394       0.001012
   17.00      0.8459       0.8456       0.000310
   17.50      0.8521       0.8516       0.000486
   18.00      0.8653       0.8573       0.008009
   18.50      0.8633       0.8628       0.000460
   19.00      0.8588       0.8681      -0.009315
   19.50      0.8725       0.8733      -0.000765
   20.00      0.8857       0.8782       0.007528
```

Table 4.2: (*Continued*)

```
ifd =   3    ncase =   3    ncall = 989
    t      u1(z=zL,t)   ua1(z=zL,t)      diff
  0.00       0.0000       0.0000       0.000000
  0.50       0.0000       0.0000       0.000000
  1.00       0.0000       0.0000       0.000000
  1.50       0.0000       0.0000       0.000000
  2.00      -0.0000       0.0000      -0.000004
  2.50      -0.0001       0.0000      -0.000071
  3.00      -0.0004       0.0000      -0.000371
  3.50       0.0036       0.0000       0.003644
  4.00       0.0129       0.0000       0.012882
  4.50      -0.0751       0.0000      -0.075105
  5.00       0.3556       0.6065      -0.250942
  5.50       0.6277       0.6214       0.006266
  6.00       0.6372       0.6357       0.001422
  6.50       0.6499       0.6495       0.000386
  7.00       0.6626       0.6628      -0.000264
  7.50       0.6756       0.6756      -0.000041
  8.00       0.6880       0.6880       0.000055
  8.50       0.6999       0.6998       0.000006
  9.00       0.7113       0.7113      -0.000008
  9.50       0.7223       0.7223       0.000004
 10.00       0.7329       0.7329       0.000004
 10.50       0.7431       0.7431       0.000002
 11.00       0.7529       0.7529       0.000003
 11.50       0.7624       0.7624       0.000002
 12.00       0.7715       0.7715       0.000003
 12.50       0.7802       0.7802       0.000002
 13.00       0.7886       0.7886       0.000002
 13.50       0.7968       0.7968       0.000003
 14.00       0.8046       0.8046       0.000002
 14.50       0.8121       0.8121       0.000002
 15.00       0.8193       0.8193       0.000002
 15.50       0.8263       0.8263       0.000002
 16.00       0.8330       0.8330       0.000002
 16.50       0.8394       0.8394       0.000002
 17.00       0.8456       0.8456       0.000002
 17.50       0.8516       0.8516       0.000002
 18.00       0.8573       0.8573       0.000002
 18.50       0.8628       0.8628       0.000002
 19.00       0.8681       0.8681       0.000002
 19.50       0.8733       0.8733       0.000002
 20.00       0.8782       0.8782       0.000002
```

Table 4.2: (*Continued*)

```
ifd =   4    ncase =   4    ncall = 2176
     t      u1(z=zL,t)   ua1(z=zL,t)      diff
   0.00       0.0000       0.0000      0.000000
   0.50      -0.0000       0.0000     -0.000000
   1.00      -0.0000       0.0000     -0.000000
   1.50      -0.0000       0.0000     -0.000000
   2.00      -0.0000       0.0000     -0.000000
   2.50      -0.0000       0.0000     -0.000000
   3.00      -0.0000       0.0000     -0.000000
   3.50      -0.0000       0.0000     -0.000000
   4.00      -0.0000       0.0000     -0.000000
   4.50       0.0011       0.0000      0.001075
   5.00       0.3330       0.6065     -0.273537
   5.50       0.6007       0.6214     -0.020714
   6.00       0.6363       0.6357      0.000568
   6.50       0.6495       0.6496     -0.000066
   7.00       0.6628       0.6628     -0.000033
   7.50       0.6756       0.6756     -0.000039
   8.00       0.6879       0.6880     -0.000043
   8.50       0.6998       0.6998     -0.000046
   9.00       0.7112       0.7113     -0.000049
   9.50       0.7222       0.7223     -0.000052
  10.00       0.7328       0.7329     -0.000055
  10.50       0.7430       0.7431     -0.000057
  11.00       0.7528       0.7529     -0.000059
  11.50       0.7623       0.7624     -0.000061
  12.00       0.7714       0.7715     -0.000063
  12.50       0.7801       0.7802     -0.000064
  13.00       0.7886       0.7886     -0.000065
  13.50       0.7967       0.7968     -0.000066
  14.00       0.8045       0.8046     -0.000067
  14.50       0.8120       0.8121     -0.000068
  15.00       0.8192       0.8193     -0.000069
  15.50       0.8262       0.8263     -0.000069
  16.00       0.8329       0.8330     -0.000069
  16.50       0.8393       0.8394     -0.000070
  17.00       0.8455       0.8456     -0.000070
  17.50       0.8515       0.8516     -0.000070
  18.00       0.8572       0.8573     -0.000070
  18.50       0.8628       0.8628     -0.000070
  19.00       0.8681       0.8682     -0.000069
  19.50       0.8732       0.8733     -0.000069
  20.00       0.8781       0.8782     -0.000069
```

Table 4.2: (*Continued*)

where p is the partial pressure of O_2 in the blood, which is given by a fractional-order polynomial

$$\alpha p^{\sigma+1} + (\beta - u_1)p^\sigma + \alpha\gamma^\sigma p - \gamma^\sigma u_1 = 0 \tag{4.4d}$$

where σ has the fractional value 2.5.

Eqs. (4.4) can be accommodated within the MOL framework. This could be done in two ways:

- A nonlinear solver, such as a variant of Newton's method, could be used at the beginning of the MOL/ODE routine pde_1 of Listing 4.1 to calculate p from eq. (4.4d) for the input vector u_1. Then, eq. (4.4c) could be used in the programming of eq. (4.4a).

- A possible computational limitation of the preceding procedure might be the use of the nonlinear solver for $n = 21$ values of u_1 each time pde_1 is called. To reduce this computational requirement, eq. (4.4d) could be solved for p for a series of values of u_1 that covers the expected range in u_1. A 1D table of p against u_1 would then be constructed, and interpolation within this table at the beginning of pde_1 would be used to obtain p. This interpolation could be by any established approach, for example, by splines, which would be faster than solving eq. (4.4d) break directly.

Eqs. (4.4) are termed a *differential algebraic equation* (DAE) system since they include both differential equations (4.4a,b) and algebraic equations (4.4c,d). DAE systems generally are not as straightforward to solve as ODE/PDEs and have been studied extensively [2]. The proposed procedures of solving the algebraic equations at the beginning of the MOL/ODE routine have been used effectively, but they are not always successful, depending on the *index* of the DAE system.

(4.5) Conclusions and summary

A system of first-order hyperbolic PDEs, eqs. (4.1), has been solved numerically by the MOL, and the solution has been evaluated against an analytical solution. Since hyperbolic PDEs can propagate steep moving fronts, special attention may have to be given to the approximation of the spatial first-order derivatives. In the preceding discussion, a comparison of FD and flux limiter approximations of the spatial derivatives was considered.

For the case of significant smoothing of the solution, the approximations generally worked well, although the effect of numerical diffusion in the case of 2pu FD was evident. For the case of limited smoothing of a steep moving front, the flux limiter was clearly superior since it minimized numerical diffusion and eliminated numerical oscillation, but did require substantially more computation to achieve this positive result. Thus, with each new problem, some experimentation with the MOL numerical approximations may be required, with careful attention to the accuracy of the numerical solutions. This assessment of accuracy generally has to be indirect and inferred since usually an analytical solution will not be available to validate the numerical solution.

References

[1] Bateman, H. (1932), *Partial Differential Equations of Mathematical Physics*, Cambridge University Press, Cambridge.

[2] Brennan, K.S., S.L. Campbell, and L.R. Petzold (1989), *Numerical Solution if Initial-Value Problems in Differential-Algebraic Equations*, North Holland, New York.

[3] Griffiths, G.W. and W.E. Schiesser (2012), *Traveling Wave Analysis of Partial Differential Equations*, Elsevier/Academic Press, Oxford.

[4] Press, W.H., S.A. Teukolsky, W.T. Vetterling, and B.P. Flannery (2007), *Numerical Recipes: The Art of Scientific Computing*, 3rd edn, Cambridge University Press, Cambridge.

[5] Sangren, W.C. and C.W. Sheppard (1953), A mathematical derivation of the exchange of a labelled substance between a liquid flowing in a vessel and an external compartment, *Bulletin of Mathematical Biophys.*, **15**, 387–394.

[6] Schiesser, W.E. (1994), *Computational Mathematics in Engineering and Applied Science: ODEs, DAEs, and PDEs*, CRC Press, Boca Raton, FL.

[7] Xie, D., R.K. Dash, and D.A. Beard (2009), An improved algorithm and its parallel implementation for solving a general blood-tissue transport and metabolism model, *Journal of Computational Physics*, **228**, 7850–7861.

5

TWO-FLUID/MEMBRANE MODEL

This chapter extends the discussion of the first-order hyperbolic partial differential equation (PDE) model in Chapter 4 by introducing a 2D, 3-PDE model for a two-fluid/membrane system. The system is depicted in Fig. 5.1.

The equations for the model are considered in the next section. Note, in particular, the three dependent variables u_1, u_2, u_3 for which the PDEs are subsequently discussed and programmed.

The intent of this chapter is to:

- Present a PDE model for a two-fluid/membrane PDE model including the required initial conditions (ICs) and boundary Conditions (BCs).
- Discuss the format of the model as a *hyperbolic-parabolic* PDE system in 2D.
- Illustrate the coding of the model through a series of R routines, including the use of library routines for integration of the PDE derivatives in time and space.
- Present the computed model solution in numerical and graphical (plotted) formats.
- Discuss the features of the numerical solution and the performance of the algorithms used to compute the solution.
- Consider extensions of the model, including nonlinear parameters and alternate BCs.

Method of Lines PDE Analysis in Biomedical Science and Engineering, First Edition. William E. Schiesser.
© 2016 John Wiley & Sons, Inc. Published 2016 by John Wiley & Sons, Inc.
Companion website: www.wiley.com/go/Schiesser/PDE_Analysis

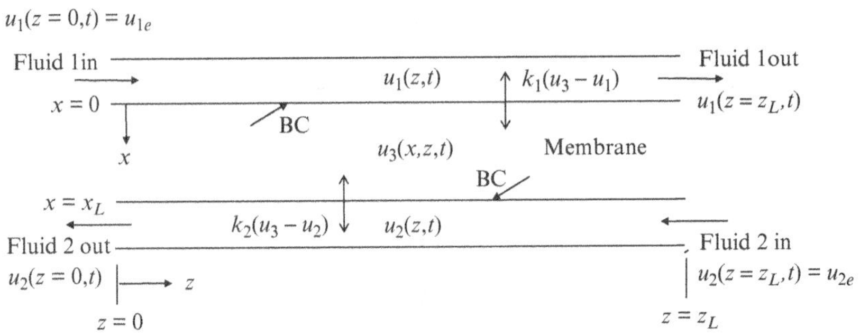

$u_1(z = 0, t) = u_{1e}$

Fluid 1 in ——————————————————————————————— Fluid 1 out

$x = 0$ $u_1(z,t)$ $\uparrow k_1(u_3 - u_1)$ $u_1(z = z_L, t)$

BC

x $u_3(x,z,t)$ Membrane

BC

$x = x_L$

$k_2(u_3 - u_2) \downarrow$ $u_2(z,t)$

Fluid 2 out ——————————————————————————————— Fluid 2 in

$u_2(z = 0, t)$ |———▸ z | $u_2(z = z_L, t) = u_{2e}$

$z = 0$ $z = z_L$

ICs: $u_1(z, t = 0) = u_{10}$ $u_2(z, t = 0) = u_{20}$ $u_3(x, z, t = 0) = u_{30}$

BCs: $D_m \dfrac{\partial u_3(x = 0, z, t)}{\partial x} = -k_1(u_1(z,t) - u_3(x = 0, z, t))$

$D_m \dfrac{\partial u_3(x = x_L, z, t)}{\partial x} = k_2(u_2(z,t) - u_3(x = x_L, z, t))$

Figure 5.1   Diagram of a two-fluid/membrane system

(5.1) 2D, 3-PDE model

The model PDEs are

$$\frac{\partial u_1}{\partial t} = -v_1 \frac{\partial u_1}{\partial z} + k_1(u_3(x = 0, z, t) - u_1) \tag{5.1a}$$

$$\frac{\partial u_2}{\partial t} = -v_2 \frac{\partial u_2}{\partial z} + k_2(u_3(x = x_L, z, t) - u_2) \tag{5.1b}$$

$$\frac{\partial u_3}{\partial t} = D_m \frac{\partial^2 u_3}{\partial x^2} \tag{5.1c}$$

where

u_1	fluid 1 concentration
u_2	fluid 2 concentration
u_3	membrane concentration
z	distance along the fluid channels
x	distance through the membrane
t	time
v_1	fluid 1 velocity
v_2	fluid 2 velocity
D_m	membrane effective diffusivity
k_1	fluid 1 to membrane mass transfer coefficient
k_2	fluid 2 to membrane mass transfer coefficient
x_L	membrane thickness
z_L	fluid channels effective length

The concentrations u_1, u_2, u_3 might represent a drug or metabolite, and the membrane might represent a porous membrane or tissue.

$u_3(x = 0, z, t)$ and $u_3(x = x_L, z, t)$ in eqs. (5.1a) and (5.1b), respectively, indicate that u_3 is a function of z as well as x. The dependency of u_3 on z and x is also reflected in the ICs and BCs and the coding that follows (even though this may not be obvious from eq. (5.1c) alone).

The initial conditions are:

$$u_1(z, t = 0) = f_1(z); \quad u_2(z, t = 0) = f_2(z); \quad u_3(x, z, t = 0) = f_3(x, z)$$
$$\text{(5.2a,b,c)}$$

where the three functions f_1, f_2, f_3 are typically constants.

The boundary conditions are:

$$u_1(z = 0, t) = g_1(t) \tag{5.3a}$$

$$u_2(z = z_L, t) = g_2(t) \tag{5.3b}$$

$$D_m \frac{\partial u_3(x = 0, z, t)}{\partial x} = -k_1(u_1(z, t) - u_3(x = 0, z, t)) \tag{5.3c}$$

$$D_m \frac{\partial u_3(x = x_L, z, t)}{\partial x} = k_2(u_2(z, t) - u_3(x = x_L, z, t)) \tag{5.3d}$$

where the two functions g_1, g_2 are typically constants. Eqs. (5.3c,d) are termed *third type* or *Robin* since they involve the dependent variables $u_3(x = 0, z, t), u_3(x = x_L, z, t)$ and their derivatives $\dfrac{\partial u_3(x = 0, z, t)}{\partial x}, \dfrac{\partial u_3(x = x_L, z, t)}{\partial x}$. Physically, eqs. (5.3c,d) equate the diffusion flux (LHS) at $x = 0, x_L$ to the mass transfer flux (RHS).

Eqs. (5.1), (5.2), and (5.3) constitute the 2D, 3-PDE model. The output that might be of particular interest includes the exiting fluid concentrations $u_1(z = z_L, t)$, $u_2(z = 0, t)$. The required parameters to be defined numerically are z_L, x_L, v_1, v_2, D_m, k_1, k_2.

(5.2) MOL analysis

The dependent variables defined by eqs. (5.1) are $u_1(z, t), u_2(z, t), u_3(x, z, t)$. Thus, the model is 1D in u_1, u_2 and 2D in u_3. The spatial grid for the MOL analysis can be summarized as shown in Table 5.1

u1(i*dz)	i index in z
	i=1,2,...,nz
u2(i*dz)	i index in z
	i=1,2,...,nz
u3(i*dz,j*dx)	i index in z
	i=1,2,...,nz
	j index in x
	j=1,2,...,nx

Table 5.1: Indexing for the MOL analysis of eqs. (5.1)

where dx, dz are the grid spacings in x, z, that is, dz $= \Delta z = (z_L - 0)/(nz - 1)$, dx $= \Delta x = (x_L - 0)/(nx - 1)$. This indexing is used in the R routines within fors.

In the following routines, nx=9, nz=21 so that the total number of ODEs programmed in the MOL/ODE routine is

$$nx*nz + nz + nz = nz*(nx + 2) = 21*(9+2) = 231$$

(5.2.1) MOL/ODE routine

The MOL/ODE routine for the programming of eqs. (5.1) and (5.3) is in Listing 5.1.

```
  pde_1=function(t,u,parms) {
#
# Function pde_1 computes the t derivative vector
# of the u vector
#
# One vector to three PDEs
  u1=rep(0,nz);u2=rep(0,nz);
  u3=matrix(0,nrow=nz,ncol=nx);
  for (i in 1:nz){
    for (j in 1:nx){
      u3[i,j]=u[(i-1)*nx+j];
    }
    u1[i]=u[nx*nz+i];
    u2[i]=u[nx*nz+nz+i];
  }
#
# Boundary conditions
  u1[1]=u1e;u2[nz]=u2e;
#
# u1z, u2z
#
#    ifd = 1: Two-point upwind finite difference (2pu)
     if(ifd==1){ u1z=dss012(0,zL,nz,u1,v1); }
     if(ifd==1){ u2z=dss012(0,zL,nz,u2,v2); }
#
#    ifd = 2: Five-point center finite difference (5pc)
     if(ifd==2){ u1z=dss004(0,zL,nz,u1); }
     if(ifd==2){ u2z=dss004(0,zL,nz,u2); }
#
#    ifd = 3: Five-point biased upwind approximation (5pbu)
     if(ifd==3){ u1z=dss020(0,zL,nz,u1,v1); }
     if(ifd==3){ u2z=dss020(0,zL,nz,u2,v2); }
#
#    ifd = 4: van Leer flux limiter
     if(ifd==4){ u1z=vanl(0,zL,nz,u1,v1); }
```

```
      if(ifd==4){ u2z=van1(0,zL,nz,u2,v2); }
#
# u3xx
    u3x=matrix(0,nrow=nz,ncol=nx);
   u3xx=matrix(0,nrow=nz,ncol=nx);
   for(i in 1:nz){
     u3x[i,1]=-(k1/Dm)*(u1[i]-u3[i,1]);
     u3x[i,nx]=(k2/Dm)*(u2[i]-u3[i,nx]);
     nl=2;nu=2;
     u3xx[i,]=dss044(0,xL,nx,u3[i,],u3x[i,],nl,nu);
   }
#
# Temporal derivatives
    u1t=rep(0,nz);u2t=rep(0,nz);
    u3t=matrix(0,nrow=nz,ncol=nx);
#
#   u1t, u2t, u3t
    for(i in 1:nz){
      if(i==1){
        u1t[i]=0;
      }else{
        u1t[i]=-v1*u1z[i]+k1*(u3[i,1]-u1[i]);
      }
      if(i==nz){
        u2t[i]=0;
      }else{
        u2t[i]=-v2*u2z[i]+k2*(u3[i,nx]-u2[i]);
      }
      for(j in 1:nx){
        u3t[i,j]=Dm*u3xx[i,j];
      }
    }
#
# Three PDE derivatives to one vector
  ut=rep(0,nx*nz+2*nz);
  for (i in 1:nz){
    for (j in 1:nx){
      ut[(i-1)*nx+j]=u3t[i,j];
    }
    ut[nx*nz+i]        =u1t[i];
    ut[nx*nz+nz+i]     =u2t[i];
  }
#
# Increment calls to pde_1
  ncall<<-ncall+1;
#
```

```
# Return derivative vector
  return(list(c(ut)));
}
```

Listing 5.1: MOL/ODE routine pde_1 for eqs. (5.1) and (5.3)

We can note the following details about Listing 5.1.

- Function pde_1 is defined. The input arguments are: (1) the current value of t; (2) the vector of dependent variables, u, of length nx*nz + nz + nz = 9*21 + 2*21 = 231 (the number of ODEs to be integrated that approximate eqs. (5.1)); and (3) the parameters passed to pde_1, which is unused.

```
  pde_1=function(t,u,parms) {
#
# Function pde_1 computes the t derivative vector
# of the u vector
```

- The dependent variable vector u is placed in two vectors, u1, u2, and one matrix, u3, to facilitate the programming of eqs. (5.1).

```
#
# One vector to three PDEs
  u1=rep(0,nz);u2=rep(0,nz);
  u3=matrix(0,nrow=nz,ncol=nx);
  for (i in 1:nz){
    for (j in 1:nx){
      u3[i,j]=u[(i-1)*nx+j];
    }
    u1[i]=u[nx*nz+i];
    u2[i]=u[nx*nz+nz+i];
  }
```

The outer for with index i steps through z and the inner for with index j steps through x. Note that only the outer for applies to u_1, u_2 since these dependent variables depend only on z while the inner and outer fors apply to u_3, which is a function of x and z (see Fig. 5.1 and Table 5.1).

- BCs (5.3c,d) are programmed.

```
#
# Boundary conditions
  u1[1]=u1e;u2[nz]=u2e;
```

u1e, u2e are entering concentrations (constants) set in the main program discussed next. Note the use of subscripts 1 and nz corresponding to $z = 0$ and $z = z_L$, respectively.

- The derivatives $\dfrac{\partial u_1}{\partial z}$ and $\dfrac{\partial u_2}{\partial z}$ in eqs. (5.1a) and (5.1b) are computed by one of four approximations selected by ifd (set in the main program to one of the values ifd=1,2,3,4).

```
#
# u1z, u2z
#
#    ifd = 1: Two point upwind finite difference (2pu)
     if(ifd==1){ u1z=dss012(0,zL,nz,u1,v1); }
     if(ifd==1){ u2z=dss012(0,zL,nz,u2,v2); }
#
#    ifd = 2: Five point center finite difference (5pc)
     if(ifd==2){ u1z=dss004(0,zL,nz,u1); }
     if(ifd==2){ u2z=dss004(0,zL,nz,u2); }
#
#    ifd = 3: Five point biased upwind approximation (5pbu)
     if(ifd==3){ u1z=dss020(0,zL,nz,u1,v1); }
     if(ifd==3){ u2z=dss020(0,zL,nz,u2,v2); }
#
#    ifd = 4: van Leer flux limiter
     if(ifd==4){ u1z=vanl(0,zL,nz,u1,v1); }
     if(ifd==4){ u2z=vanl(0,zL,nz,u2,v2); }
```

Test runs for the four approximations indicate that the van Leer flux limiter should be used (ifd=4) to eliminate numerical diffusion and oscillation in the numerical solution (see also the discussion in Chapter 4).

- The derivative $\dfrac{\partial^2 u_3}{\partial x^2}$ in eq. (5.1c) is computed by the spatial differentiator dss044.

```
#
# u3xx
   u3x=matrix(0,nrow=nz,ncol=nx);
   u3xx=matrix(0,nrow=nz,ncol=nx);
   for(i in 1:nz){
     u3x[i,1]=-(k1/Dm)*(u1[i]-u3[i,1]);
     u3x[i,nx]=(k2/Dm)*(u2[i]-u3[i,nx]);
     nl=2;nu=2;
     u3xx[i,]=dss044(0,xL,nx,u3[i,],u3x[i,],nl,nu);
   }
```

The array u3x for $\dfrac{\partial u_3}{\partial x}$ is used by dss044 only at the boundaries $x = 0, x_L$. In other words, u3x includes BCs (5.3c,d) (for u3x[i,1], u3x[i,nx]). nl=nu=2 indicates that the BCs are *Neumann* since the derivatives at the boundaries $x = 0, x_L$ are specified.

Vectorization is used with [i,], that is, a particular value of i (z) for all values of x (since the computed derivative is with respect to x).

Thus, eq. (5.1c) is integrated nz times (i=1,2,...,nz) in each call to pde_1. This demonstrates that the total number of ODEs, nx*nz + 2*nz, increases rapidly with an increase in the number of grid points due to the product nx*nz, and care must be given to selecting the values of nx and nz to ensure that the total number of ODEs is tractable, but that convergence of the numerical solution to reasonable accuracy is achieved.

Since eqs. (5.1a) and (5.1b) are hyperbolic in z, nz=21 was selected to give resolution of fronts moving in z. Eq. (5.1c) is parabolic in x so that smoothing of the solution occurs through diffusion and therefore fewer points were used (nx=9). However, the effect of these values on the solution should be studied (h refinement) to give some assurance of convergence of the solution to an acceptable accuracy.

- Two vectors for the derivatives in t of u_1, u_2, and a matrix for the derivative in t of u_3 are declared.

```
#
# Temporal derivatives
    u1t=rep(0,nz);u2t=rep(0,nz);
    u3t=matrix(0,nrow=nz,ncol=nx);
```

- Derivatives u1t $= \dfrac{\partial u_1}{\partial t}$, u2t $= \dfrac{\partial u_2}{\partial t}$, and u3t $= \dfrac{\partial u_3}{\partial t}$ are computed according to eqs. (5.1).

```
#
#    u1t, u2t, u3t
    for(i in 1:nz){
      if(i==1){
        u1t[i]=0;
      }else{
        u1t[i]=-v1*u1z[i]+k1*(u3[i,1]-u1[i]);
      }
      if(i==nz){
        u2t[i]=0;
      }else{
        u2t[i]=-v2*u2z[i]+k2*(u3[i,nx]-u2[i]);
      }
      for(j in 1:nx){
        u3t[i,j]=Dm*u3xx[i,j];
      }
    }
```

Since BCs (5.3a) and (5.3b) specify constants, their derivatives are zero, u1t[1]=0, u2t[nz]=0. The close resemblance of the coding and the PDEs (eqs. (5.1)) is an important feature of the MOL.

- All of the derivatives in t have been computed so they are placed in a vector ut with two nested fors for use by the ODE integrator (ode called in the main program discussed next).

```
#
```

```
# Three PDE derivatives to one vector
  ut=rep(0,nx*nz+2*nz);
  for (i in 1:nz){
    for (j in 1:nx){
      ut[(i-1)*nx+j]=u3t[i,j];
    }
    ut[nx*nz+i]        =u1t[i];
    ut[nx*nz+nz+i]     =u2t[i];
  }
```

Note that the subscripting is the reverse of that used for u at the beginning of pde_1 to ensure that each dependent variable in u has a derivative placed in the corresponding position in ut. In other words, this positioning is required so that each derivative, when integrated numerically, produces the corresponding dependent variable.

- The number of calls to pde_1 is incremented and returned to the main program with <<-.

```
#
# Increment calls to pde_1
  ncall<<-ncall+1;
```

- The derivative vector ut is returned to the ODE integrator.

```
#
# Return derivative vector
  return(list(c(ut)));
}
```

c is the R vector operator, and the vector is returned as a list as required by the R ODE integrators. The final } concludes pde_1.

The main program that calls pde_1 is considered next.

(5.2.2) Main program

The main program for the model of eqs. (5.1) to (5.3) follows.

```
#
# Delete previous workspaces
  rm(list=ls(all=TRUE))
#
#    2D, 3-PDE fluid-membrane transport model
#
# Access ODE integrator
  library("deSolve");
#
# Access files
  setwd("g:/chap5");
```

```
  source("pde_1.R");source("dss044.R");
  source("dss004.R");source("dss012.R");
  source("dss020.R");source("vanl.R")   ;
#
# Step through cases
  for(ncase in 1:1){
#
# Model parameters
  u10=0; u20=0; u30=0;
   nx=9; nz=21;
   v1=1; v2=-1;
   k1=1;  k2=1; Dm=0.1;
   zL=5;  xL=1;
  u1e=1; u2e=0;
#
# Select an approximation for the convective derivatives
# u1z, u2z
#
#   ifd = 1: Two-point upwind approximation
#
#   ifd = 2: Centered approximation
#
#   ifd = 3: Five-point, biased upwind approximation
#
#   ifd = 4: van Leer flux limiter
#
  ifd=4;
#
# Level of output
#
#   Detailed output   - ip = 1
#
#   Brief (IC) output - ip = 2
#
  ip=1;
#
# Initial condition
  u0=rep(0,nz*(nx+2));
  for(i in 1:nz){
    for(j in 1:nx){
      u0[(i-1)*nx+j]=u30;
#     cat(sprintf("\n i=%2d j=%2d (i-1)*nx+j=%3d",
#                  i,j,(i-1)*nx+j));
    }
    u0[nx*nz+i]    =u10;
    u0[nx*nz+nz+i] =u20;
```

```
  }
nrow(u0)
ncol(u0)
#
# Grid in t
  t0=0;tf=25;nout=51;
  tout=seq(from=t0,to=tf,by=(tf-t0)/(nout-1));
  ncall=0;
#
# ODE integration
  out=ode(func=pde_1,times=tout,y=u0);
#
# Store solution
  u1=matrix(0,nrow=nout,ncol=nz);
  u2=matrix(0,nrow=nout,ncol=nz);
   t=rep(0,nout);
  for(it in 1:nout){
    for(i in 1:nz){
      u1[it,i]=out[it,nx*nz+i+1];
      u2[it,i]=out[it,nx*nz+nz+i+1];
    }
    t[it]=out[it,1];
  }
#
# Display ifd, ncase, ncall
  cat(sprintf("\n\n ifd = %2d ncase = %2d ncall = %2d",
              ifd,ncase,ncall));
#
# Display numerical solution
  if(ip==1){
    cat(sprintf(
      "\n\n      t      u1(z=zL,t)  u2(z=0,t)\n"));
    for(it in 1:nout){
      cat(sprintf(
        "%7.2f%12.4f%12.4f\n",t[it],u1[it,nz],u2[it,1]));
    }
  }
#
# Plot u1(z=zL,t), u2(z=0,t)
#
# u1(z=zL,t)
  par(mfrow=c(1,1));
  plot(tout,u1[,nz],type="l",xlab="t",ylab="u1(z=zL,t)",
    main="u1(z=zL,t) vs t",col="black",lwd=2,lty=1);
#
# u2(z=0,t)
```

```
   par(mfrow=c(1,1));
   plot(tout,u2[,1],type="l",xlab="t",ylab="u2(z=0,t)",
     main="u2(z=0,t) vs t",col="black",lwd=2,lty=1);
#
# Next case
   }
```

Listing 5.2: Main program for eqs. (5.1), (5.2), and (5.3)

We can observe the following details in Listing 5.2.

- Previous workspaces are removed.

```
#
# Delete previous workspaces
   rm(list=ls(all=TRUE))
#
#    2D, 3-PDE fluid-membrane transport model
```

- The R ODE library deSolve is accessed, and the routines required for the MOL solution of eqs. (5.1) are referenced with the source utility.

```
#
# Access ODE integrator
   library("deSolve");
#
# Access files
   setwd("g:/chap5");
   source("pde_1.R") ;source("dss044.R");
   source("dss004.R");source("dss012.R");
   source("dss020.R");source("van1.R")   ;
```

The setwd (set working directory) requires editing for the local computer (note the use of / rather than the usual \).

- One case is programmed, but the for can be used to program multiple cases, e.g., for parameter variation as in Chapter 4.

```
#
# Step through cases
   for(ncase in 1:1){
```

- The model parameters are defined numerically.

```
#
# Model parameters
   u10=0; u20=0; u30=0;
     nx=9; nz=21;
```

```
v1=1; v2=-1;
k1=1;   k2=1; Dm=0.1;
zL=5    xL=1;
u1e=1; u2e=0;
```

Note in particular that $u_1(z = 0, t) =$ u1e=1 moves the solution away from the homogeneous ICs of eqs. (5.2) (u10 = u20 = u30 = 0). That is, a unit step is imposed at the left boundary $z = 0$.

- Four approximations for the derivatives $\dfrac{\partial u_1}{\partial z}$, $\dfrac{\partial u_2}{\partial z}$, in eqs. (5.1a), (5.1b) can be selected.

```
#
# Select an approximation for the convective derivatives
# u1z, u2z
#
#    ifd = 1: Two point upwind approximation
#
#    ifd = 2: Centered approximation
#
#    ifd = 3: Five point, biased upwind approximation
#
#    ifd = 4: van Leer flux limiter
#
   ifd=4;
```

The van Leer flux limiter is selected for resolution of moving fronts in z in accordance with the second case in Chapter 4.

- Detailed numerical output is selected that will be reviewed subsequently.

```
#
# Level of output
#
#    Detailed output   - ip = 1
#
#    Brief (IC) output - ip = 2
#
   ip=1;
```

- ICs (5.2) are defined numerically using u10 = u20 = u30 to define an IC vector u0 of length nx*nz+2*nz. First, IC (5.2c) for $u_3(x, z, t = 0)$ is used in two nested fors to step through x and z for a total of nx*nz = 9*21 = 189 ODEs. Then, $u_1(z, t = 0)$ and $u_2(z, t = 0)$ are defined for a total of 2*nz = 2*21 = 42 additional ODEs, for a total of 189 + 42 = 231 ODEs.

```
#
# Initial condition
   u0=rep(0,nz*(nx+2));
```

```
    for(i in 1:nz){
      for(j in 1:nx){
        u0[(i-1)*nx+j]=u30;
#       cat(sprintf("\n i=%2d    j=%2d   (i-1)*nx+j=%3d",
#                     i,j,(i-1)*nx+j));
      }
      u0[nx*nz+i]      =u10;
      u0[nx*nz+nz+i]  =u20;
    }
nrow(u0)
ncol(u0)
```

The output statement was used to check the indexing in the two nested fors (it can be used by deleting the #). nrow and ncol are used to confirm the expected dimensions of u0 (a 231-vector).

- The grid in t for $0 \leq t \leq 25$ with 51 points (including $t = 0$) is defined so that tout = 0,0.5,...,25.

```
#
# Grid in t
  t0=0;tf=25;nout=51;
  tout=seq(from=t0,to=tf,by=(tf-t0)/(nout-1));
  ncall=0;
```

The counter for the number of calls to the ODE routine pde_1 (discussed next) is also initialized.

- The 231-ODE system is integrated by ode.

```
#
# ODE integration
  out=ode(func=pde_1,times=tout,y=u0);
```

The arguments to ode are: (1) the MOL/ODE routine pde_1 of Listing 5.1, (2) the vector of output values of t, tout, and (3) the IC vector u0. func,times,y are reserved names. The numerical solution is returned in matrix out. The length of u0 (231) informs ode of the number of ODEs to be integrated.

- The numerical solution for $u_1(z,t)$ and $u_2(z,t)$ in out is placed in two matrices, u1,u2, for subsequent display numerically and graphically. The offset 1 in nx*nz+i+1,, nx*nz+nz+i+1 is used since the first position in out is for t, that is, t[it]=out[it,1].

```
#
# Store solution
  u1=matrix(0,nrow=nout,ncol=nz);
  u2=matrix(0,nrow=nout,ncol=nz);
  t=rep(0,nout);
```

```
      for(it in 1:nout){
        for(i in 1:nz){
          u1[it,i]=out[it,nx*nz+i+1];
          u2[it,i]=out[it,nx*nz+nz+i+1];
        }
      }
```

- The counter for the calls to pde_1 of Listing 5.2 is displayed as a measure of the computational effort required to compute the numerical solution.

```
#
# Display ifd, ncase, ncall
  cat(sprintf("\n\n ifd = %2d   ncase = %2d   ncall = %2d",
              ifd,ncase,ncall));
```

- The concentrations of the exit streams, $u_1(z = z_L, t)$ and $u_2(z = 0, t)$, are displayed as a function of t.

```
#
# Display numerical solution
  if(ip==1){
    cat(sprintf(
      "\n\n     t       u1(z=zL,t)  u2(z=0,t)\n"));
    for(it in 1:nout){
      cat(sprintf(
        "%7.2f%12.4f%12.4f\n",t[it],u1[it,nz],u2[it,1]));
    }
  }
```

Note the use of subscripts 1,nz corresponding to $z = 0, z_L$.

- The concentrations of the exit streams, $u_1(z = z_L, t)$ and $u_2(z = 0, t)$, are plotted. Vectorization as u1[,nz],u2[,1] is used for all values of t at $z = z_L, z = 0$. The utility matplot accepts these matrices as vectors. As might be expected from the name, it will also accept the matrices u1,u2 (without vectorization) and produce plots of $u_1(z, t)$ and $u_2(z, t)$ against t with z as a parameter (according to the second subscript).

```
#
# Plot u1(z=zL,t), u2(z=0,t)
#
# u1(z=zL,t)
  par(mfrow=c(1,1));
  matplot(tout,u1[,nz],type="l",xlab="t",ylab="u1(z=zL,t)",
    main="u1(z=zL,t) vs t",col="black",lwd=2,lty=1);
#
# u3(z=0,t)
```

```
    par(mfrow=c(1,1));
    matplot(tout,u2[,1],type="l",xlab="t",ylab="u2(z=0,t)",
      main="u2(z=0,t) vs t",col="black",lwd=2,lty=1);
#
# Next case
  }
```

The final } concludes the for in ncase (with just one value, ncase=1).

This completes the programming of eqs. (5.1)–(5.3). The numerical and graphical outputs are considered next.

(5.3) Model output

The output from the main program is given in Table 5.2 and Figs. 5.2, 5.3.
   We can note the following details about the output in Table 5.2 and Figs. 5.2, 5.3.

- $u_2(z = 0, t)$ responds almost immediately to $u_1(z = 0, t) = u_{1e} = 1$ (see BC (5.3a) in pde_1 of Listing 5.1), while $u_1(z = z_L, t)$ does not respond until $t = 4.50$

t	u1(z=zL,t)	u2(z=0,t)
0.00	0.0000	0.0000
0.50	−0.0000	0.0001
1.00	−0.0000	0.0018
1.50	−0.0000	0.0069
2.00	−0.0000	0.0149
2.50	−0.0000	0.0248
3.00	−0.0000	0.0361
3.50	−0.0000	0.0481
4.00	−0.0000	0.0606
4.50	0.0112	0.0731
5.00	0.1092	0.0856

This is to be expected since $u_2(z = 0, t)$ immediately "sees" $u_1(z = 0, t) = u_{1e} = 1$ through the membrane at $z = 0$, while the unit step must move through the system to $z = z_L$ before it appears in $u_1(z = z_L, t)$.
- Fig. 5.2 confirms this explanation since $u_1(z = z_L, t)$ takes a step at approximately $t = z_L/v = 5/1 = 5$.
- Fig. 5.3 also confirms this explanation since $u_2(z = 0, t)$ responds almost immediately to the entering $u_1(z = 0, t) = u_{1e} = 1$.
- Since the van Leer flux limiter was used, the solutions in Figs. 5.2 and 5.3 do not exhibit any numerical oscillation. Presumably, they also have small numerical diffusion (and only problem diffusion from the mass transport across the membrane), as demonstrated in Chapter 4. An analytical solution is not available to confirm this conclusion. But runs with ifd=1 demonstrate numerical diffusion and with ifd=2,3 numerical oscillation, as discussed in Chapter 4.

```
ifd =   4    ncase =   1    ncall = 3214
      t      u1(z=zL,t)   u2(z=0,t)
    0.00       0.0000       0.0000
    0.50      -0.0000       0.0001
    1.00      -0.0000       0.0018
    1.50      -0.0000       0.0069
    2.00      -0.0000       0.0149
    2.50      -0.0000       0.0248
    3.00      -0.0000       0.0361
    3.50      -0.0000       0.0481
    4.00      -0.0000       0.0606
    4.50       0.0112       0.0731
    5.00       0.1092       0.0856
    5.50       0.2404       0.0979
    6.00       0.3481       0.1101
    6.50       0.4239       0.1219
    7.00       0.4776       0.1334
    7.50       0.5176       0.1447
    8.00       0.5487       0.1556
    8.50       0.5734       0.1662
    9.00       0.5932       0.1766
    9.50       0.6093       0.1866
   10.00       0.6224       0.1964
   10.50       0.6333       0.2059
   11.00       0.6424       0.2151
   11.50       0.6500       0.2238
   12.00       0.6565       0.2320
   12.50       0.6621       0.2395
   13.00       0.6669       0.2462
   13.50       0.6712       0.2523
   14.00       0.6749       0.2576
   14.50       0.6782       0.2623
   15.00       0.6811       0.2664
   15.50       0.6838       0.2700
   16.00       0.6862       0.2731
   16.50       0.6883       0.2757
   17.00       0.6902       0.2781
   17.50       0.6919       0.2801
   18.00       0.6935       0.2818
   18.50       0.6949       0.2833
   19.00       0.6961       0.2846
   19.50       0.6973       0.2858
   20.00       0.6983       0.2868
   20.50       0.6991       0.2876
```

Table 5.2: Numerical solution for eqs. (5.1) to (5.3)

21.00	0.6999	0.2884
21.50	0.7006	0.2891
22.00	0.7013	0.2897
22.50	0.7018	0.2902
23.00	0.7023	0.2906
23.50	0.7027	0.2910
24.00	0.7031	0.2914
24.50	0.7034	0.2917
25.00	0.7037	0.2920

Table 5.2: (*Continued*)

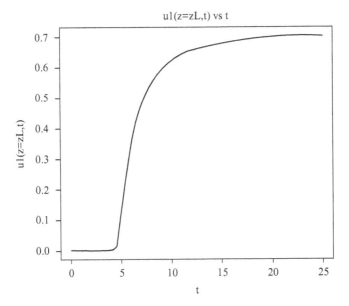

Figure 5.2   Numerical solution $u_1(z = z_L, t)$

- The computational effort is modest, `ncall` = 3214.

(5.4) Summary and conclusions

In this chapter, a 2D 3-PDE model was considered for transport between two fluid streams (1D 2-PDEs) and a membrane (2D 1-PDE). The coding in Listings 5.1 and 5.2 demonstrated a procedure for three interconnected domains, and in particular, the subscripting to accommodate this arrangement within the MOL framework. Mathematically, the model can be classified as *hyperbolic–parabolic*. Physically, it might represent

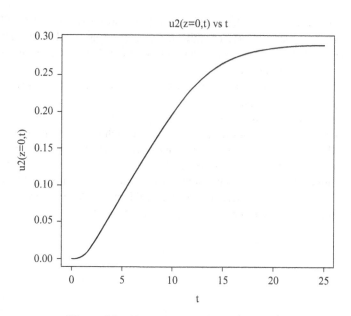

Figure 5.3 Numerical solution $u_2(z = 0, t)$

a hemodialyzer with detailed analysis of the membrane. Physiologically, it might represent a kidney or a liver. But the methodology is general and is applicable to multi-PDE, multidomain systems, including the BCs that connect the domains.

The model can be extended to include additional phenomena. For example, diffusion in the membrane in the z direction could be included, that is, eq. (5.1c) could be extended to

$$\frac{\partial u_3}{\partial t} = D_m \frac{\partial^2 u_3}{\partial x^2} + D_m \frac{\partial^2 u_3}{\partial z^2} \tag{5.1d}$$

The second derivative in z requires two BCs such as two homogeneous Neumann BCs

$$\frac{\partial u_3(x, z = 0, t)}{\partial z} = 0; \quad \frac{\partial u_3(x, z = z_L, t)}{\partial z} = 0 \tag{5.3e,f}$$

Eqs. (5.3e,f) are *no flux* BCs indicating that no diffusion occurs through the membrane boundaries at $z = 0, z_L$.

This addition of diffusion in the z direction would probably have little effect on the numerical solution since the membrane is thin ($x_L = 1$, $z_L = 5$), and therefore, diffusion is primarily in the x direction. However, this conclusion could easily be checked by comparing the solutions with and without z diffusion.

Another possibility would be to add a volumetric reaction term to eqs. (5.1) to (5.3) to model a bioreactor. This would possibly lead to various chemical species such as reactants and products, each of which would require a set of PDEs. Thus, model extensions can become relatively complicated (for example, nonlinear), but the use of computer-based analysis does not, at least in principle, have inherent limitations that would preclude the model development.

This type of analysis and reasoning illustrates how experimentation with the model structure is possible, for example, by adding terms to the equations, or even adding additional equations. In other words, modeling is not a unique process and evolves to include the principal phenomena of the physical/chemical system that should be included in the model. This process of model formulation will be directed by the experience and insights of the analyst and as new information becomes available, e.g., experimental data. The final result will, ideally, be a model that reflects the important features and characteristics of the problem system in sufficient detail and with acceptable accuracy for the intended purpose of the study.

6

LIVER SUPPORT SYSTEMS

This chapter pertains to an ordinary/partial differential equation (ODE/PDE) mathematical model for an artificial liver support system (ALSS) to remove toxins from blood as performed naturally by the liver. The basic concepts of the ALSS flow configuration are described in the following excerpt from Annesni et al. [1].

> The liver can be considered a complex large-scale biochemical reactor, because it occupies a central position in the metabolic processes. Although individual pathway for synthesis and breakdown of carbohydrates, lipids, amino acids, proteins and nucleic acids can be identified in other mammalian cells, only the liver performs all of these biochemical transformations simultaneously and is able to combine them to accomplish its vital biological task. Therefore, this organ displays a unique biological complexity; when it fails, functional replacements represent one of the most difficult challenges in substitutive medicine.

The intent of this chapter is to:

- Present an ODE/PDE model for an ALSS, including the required initial conditions (ICs) and boundary conditions (BCs).
- Discuss the model in two steps: (a) a 2-ODE patient model and (b) a 1D 8-PDE model for the extracorporeal ALSS.
- Explain the simultaneous ODEs and PDEs (coupling) of (a) and (b) above to construct a complete ODE/PDE model for the patient ALSS, programmed in a series of R routines.
- Present the computed model solution in numerical and graphical (plotted) format.

Method of Lines PDE Analysis in Biomedical Science and Engineering, First Edition. William E. Schiesser.
© 2016 John Wiley & Sons, Inc. Published 2016 by John Wiley & Sons, Inc.
Companion website: www.wiley.com/go/Schiesser/PDE_Analysis

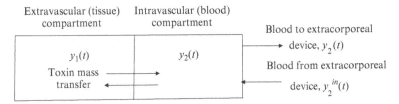

Figure 6.1   ODE patient model

- Summarize the numerical methods for the hyperbolic (convective) and hyperbolic-parabolic (convective-diffusive) PDEs in the model.

(6.1) 2-ODE patient model

The patient ODE model is illustrated in Fig. 6.1.

The ODE model is[1]

$$V_E \frac{dy_1}{dt} = K_P S(y_2^u - K_e y_1) + V_E Q_E \tag{6.1a}$$

$$V_P \frac{dy_2}{dt} = Q_B(y_2^{in} - y_2) - K_P S(y_2^u - K_e y_1) \tag{6.1b}$$

where

y_1	extravascular unbound toxin concentration
y_2	intravascular total toxin concentration
y_2^u	intravascular unbound toxin concentration
t	time
V_E	volume of extravascular compartment
V_P	volume of intravascular compartment
K_e	equilibrium ratio of unbound toxin concentrations in two compartments
K_P	mass transfer coefficient for unbound toxin between two compartments
S	mass transfer surface
Q_B	blood flow rate sent to the extracorporeal ALSS device
y_2^{in}	total toxin concentration from the ALSS device
Q_E	volumetric generation rate of unbound toxin in the extravascular compartment

Eq. (6.1a) is a balance for the unbound toxin in the extravascular (tissue) compartment. Eq. (6.1b) is a balance for the total toxin in the intravascular (blood) compartment. In particular, the rate of transfer of unbound toxin between the two compartments is given

---

[1] The ODE/PDE notation is **y** for the vector of ODE dependent variables, e.g., dy_i/dt, and **u** for the vector of PDE dependent variables, e.g., $\partial u_i/\partial t$, $\partial u_i/\partial x$.

by $\pm K_P S(y_2^u - K_e y_1)$. The nonhomogeneous term Q_E in eq. (6.1a) drives the entire model away from the initial conditions (ICs), as stated next in eqs. (6.3).

Minor rearrangement of eqs. (6.1) gives ([1], p 395)

$$\frac{dy_1}{dt} = \frac{V_P}{V_E} \frac{K_P S}{V_P} y_2^u - \frac{K_P S K_e}{V_E} y_1 + Q_E \qquad (6.2a)$$

$$\frac{dy_2}{dt} = \frac{Q_B}{V_P}(y_2^{in} - y_2) - \frac{K_P S}{V_P} y_2^u + \frac{V_E}{V_P} \frac{K_P S K_e}{V_E} y_1 \qquad (6.2b)$$

In accordance with [1], p 393, we take in eqs. (6.2) $V_P/V_E = 3/5$, $\lambda_{12} = \dfrac{K_P S}{V_P}$, $\lambda_{21} = \dfrac{K_P S K_e}{V_E}$ where $\lambda_{12}, \lambda_{21}$ are fractional transfer rates between the extravascular and intravascular compartments.

Eqs. (6.2) each require an IC.

$$y_1(t = 0) = y_{10}; \; y_2(t = 0) = y_{20} \qquad (6.3a,b)$$

y_{10}, y_{20} are constants to be specified.

Eqs. (6.2) and (6.3), along with numerical values of the parameters, constitute the ODE model for the patient. They are programmed in the following routines (which are subsequently added to the PDE model for the ALSS membrane and adsorption units).

(6.2) Patient ODE model routines

The R routines for eqs. (6.2) and (6.3) are the main program in Listing 6.1 and the ODE routine in Listing 6.2.

(6.2.1) Main program

The main program for eqs. (6.2) and (6.3) follows.

```
#
# Delete previous workspaces
  rm(list=ls(all=TRUE))
#
# Two-ODE patient model
#
# Access ODE integrator
  library("deSolve");
#
# Access functions for numerical solution
  setwd("g:/chap6");
  source("ode_1.R");
#
```

```
# Level of output
#
#   ip = 1 - graphical (plotted) solutions
#              (y1(t), y2(t) only)
#
#   ip = 2 - numerical and graphical solutions
#
  ip=2;
#
# Parameters
   VP=5000; VE=(5/3)*VP; lam12=0.0047; lam21=0.0017;  QB=150;
  QE=0.005;      alpha=1;    y2in=0.01;            y10=1;   y20=1;
  lam21=0.017;
  cat(sprintf("\n  VP = %4.1f  VE = %4.1f\n",VP,VE));
  cat(sprintf("\n  lam12 = %8.5f  lam21 = %8.5f\n",lam12,lam21));
  cat(sprintf("\n  QB = %4.1f  QE = %8.5f  alpha = %5.2f\n",
             QB,QE,alpha));
  cat(sprintf("\n  y2in = %5.3f  y10 = %5.3f  y20 = %5.3f\n",
             y2in,y10,y20));
#
# Independent variable for ODE integration
  nout=37;t0=0;tf=360;
  tout=seq(from=t0,to=tf,by=(tf-t0)/(nout-1));
#
# Initial conditions
  y0=rep(0,2);
  y0[1]=y10;
  y0[2]=y20;
  ncall=0;
#
# ODE integration
  out=lsodes(func=ode_1,y=y0,times=tout,parms=NULL)
#
# Save and display numerical solution
  y1=rep(0,nout);y2=rep(0,nout);
  for(it in 1:nout){
    y1[it]=out[it,2];y2[it]=out[it,3];
  }
  if(ip==2){
    for(it in 1:nout){
      if(it==1){
        cat(sprintf("\n       t       y1(t)      y2(t)\n"));
      }
    cat(sprintf("%7.1f%10.4f%10.4f\n",tout[it],y1[it],y2[it]));
    }
  }
```

```
#
# Calls to ODE routine
  cat(sprintf("\n\n  ncall = %3d\n\n",ncall));
#
# Plot y1, y2
  par(mfrow=c(1,1))
  plot(tout,y1,
     xlab="t (min)",ylab="y1,y2 (micromol/ml)",
     main="y1(t),y2(t)",type="l",lwd=2,
     xlim=c(0,400),ylim=c(0.5,1));
  points(tout,y1, pch="1",lwd=2);
   lines(tout,y2,type="l",lwd=2);
  points(tout,y2, pch="2",lwd=2);
```

Listing 6.1: Main program for eqs. (6.2) and (6.3)

We can note the following details of Listing 6.1.

- Previous workspaces are removed.

```
#
# Delete previous workspaces
  rm(list=ls(all=TRUE))
#
# Two-ODE patient model
```

- The library for the ODE integrator and the file with the programming of the ODEs is accessed. In the setwd (set working directory), / is used in place of the usual \.

```
#
# Access ODE integrator
  library("deSolve");
#
# Access functions for numerical solution
  setwd("g:/chap6");
  source("ode_1.R");
```

- A level of output is selected. For ip=2, detailed numerical output is produced as well as graphical (plotted) output.

```
#
# Level of output
#
#    ip = 1 - graphical (plotted) solutions
#             (y1(x,t), y2(x,t) only)
#
```

```
#     ip = 2 - numerical and graphical solutions
#
    ip=2;
```

- The parameters in eqs. (6.2) and (6.3) are defined numerically (some values are from [1]). The units of the ODE dependent variables, y_1, y_2, and parameters are listed below.

  - VP,VE: ml (milliliters)
  - QB: ml/min
  - QE: μmol/min-ml (toxin)
  - y2in,y10,y20: μmol/ml (toxin)
  - y_1, y_2 (solutions to eqs. (6.2)): μmol/ml (toxin)

```
#
# Parameters
    VP=5000; VE=(5/3)*VP; lam12=0.0047; lam21=0.0017;  QB=150;
    QE=0.005;     alpha=1;    y2in=0.01;           y10=1;   y20=1;
    lam21=0.017;
    cat(sprintf("\n  VP = %4.1f   VE = %4.1f\n",VP,VE));
    cat(sprintf("\n  lam12 = %8.5f   lam21 = %8.5f\n",lam12,
              lam21));
    cat(sprintf("\n  QB = %4.1f   QE = %8.5f   alpha = %5.2f\n",
              QB,QE,alpha));
    cat(sprintf("\n  y2in = %5.3f   y10 = %5.3f   y20 = %5.3f\n",
              y2in,y10,y20));
```

The value of lam21 is changed from 0.0017 to 0.017 to give a more pronounced transfer of toxin between the extravascular (tissue) and intravascular (blood) volumes.

- The solution interval in t is defined as $0 \leq t \leq 360$ min by the seq operator with 37 output values corresponding to $t = 0, 10,...,360$ min.

```
#
# Independent variable for ODE integration
    nout=37;t0=0;tf=360;
    tout=seq(from=t0,to=tf,by=(tf-t0)/(nout-1));
```

- ICs (6.3) are placed in a vector y0 declared (preallocated) by the rep operator for use by the ODE integrator.

```
#
# Initial conditions
    y0=rep(0,2);
    y0[1]=y10;
    y0[2]=y20;
    ncall=0;
```

The counter for calls to the ODE routine is initialized.

- The 2×2 ODE system, eqs. (6.2) and (6.3), is integrated by lsodes. As expected, the ODE routine, ode_1, the IC vector, y0, and the vector of output times, tout, are inputs to lsodes. func,y,times are reserved names. parm for passing parameters to lsodes is unused.

```
#
# ODE integration
  out=lsodes(func=ode_1,y=y0,times=tout,parms=NULL)
```

The length of y0 informs lsodes of the number of ODEs to be integrated (in this case, two).

- The ODE solutions returned by lsodes in out are placed in two vectors, y1,y2, then displayed

```
#
# Save and display numerical solution
  y1=rep(0,nout);y2=rep(0,nout);
  for(it in 1:nout){
    y1[it]=out[it,2];y2[it]=out[it,3];
  }
  if(ip==2){
    for(it in 1:nout){
      if(it==1){
        cat(sprintf("\n      t      y1(t)      y2(t)\n"));
      }
    cat(sprintf("%7.1f%10.4f%10.4f\n",tout[it],y1[it],
                y2[it]));
    }
  }
```

The offset 1 in y1[it]=out[it,2], y2[it]=out[it,3] is used since the first position in out is for t, that is, tout[it]=out[it,1].

- The number of calls to ode_1 is displayed as a measure of the computational effort required to compute the complete solution to eqs. (6.2).

```
#
# Calls to ODE routine
  cat(sprintf("\n\n  ncall = %3d\n\n",ncall));
```

- $y_1(t), y_2(t)$ are plotted against t with the utilities par,plot,points,lines. The solution curves are identified with pch="1",pch="2" as demonstrated in Fig. 6.2. Scaling of the x and y axes is defined with xlim=c(0,400), ylim=c(0.5,1) to improve the appearance of the plot (rather than using the automatic scaling from plot).

```
#
# Plot y1, y2
```

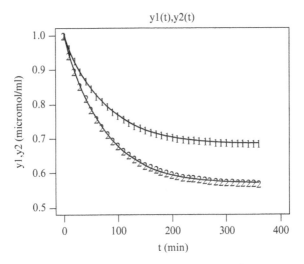

Figure 6.2 Solution of eqs. (6.2) and (6.3)

```
par(mfrow=c(1,1))
plot(tout,y1,
    xlab="t (min)",ylab="y1,y2 (micromol/ml)",
    main="y1(t),y2(t)",type="l",lwd=2,
    xlim=c(0,400),ylim=c(0.5,1));
  points(tout,y1, pch="1",lwd=2);
   lines(tout,y2,type="l",lwd=2);
  points(tout,y2, pch="2",lwd=2);
```

The ODE routine, ode_1, called by lsodes is considered next.

(6.2.2) ODE routine

ode_1 for eqs. (6.2) is in Listing 6.2.

```
  ode_1=function(t,y,parms){
#
# Function ode_1 computes the t derivative vector of
# y1(t), y2(t)
#
# One vector to two scalars
  y1=y[1];
  y2=y[2];
#
# ODEs
  y2u=alpha*y2;
  y1t=(VP/VE)*lam12*y2u-lam21*y1+QE;
```

```
  y2t=(QB/VP)*(y2in-y2)-lam12*y2u+(VE/VP)*lam21*y1;
#
# Two scalars to one vector
  yt=rep(0,2);
  yt[1]=y1t;
  yt[2]=y2t;
#
# Increment calls to ode_1
  ncall <<- ncall+1;
#
# Return derivative vector
  return(list(c(yt)));
  }
```

Listing 6.2: ODE routine for eqs. (6.2)

We can note the following details of Listing 6.2.

- The function is defined. The input argument t is the current value of t in eqs. (6.2). y is a 2-vector with $y_1(t), y_2(t)$. parms for passing parameters to ode_1 is unused, but must be included in the argument list.

  ```
    ode_1=function(t,y,parms){
  #
  # Function ode_1 computes the t derivative vector of
  # y1(t), y2(t)
  ```

- y is placed in two scalars to facilitate programming with problem oriented variables.

  ```
  #
  # One vector to two scalars
    y1=y[1];
    y2=y[2];
  ```

- Eqs. (6.2) are programmed. The parameters alpha,VP,...,lam21 are available to ode_1 from the main program (Listing 6.1) without any special designation (a feature of R).

  ```
  #
  # ODEs
    y2u=alpha*y2;
    y1t=(VP/VE)*lam12*y2u-lam21*y1+QE;
    y2t=(QB/VP)*(y2in-y2)-lam12*y2u+(VE/VP)*lam21*y1;
  ```

- The derivatives $dy_1/dt, dy_2/dt$ are placed in a 2-vector yt to return to lsodes.

```
#
# Two scalars to one vector
  yt=rep(0,2);
  yt[1]=y1t;
  yt[2]=y2t;
```

- The counter for the calls to ode_1 is incremented and returned to the main program with <<-.

```
#
# Increment calls to ode_1
  ncall <<- ncall+1;
```

- The derivative vector yt is returned to lsodes with c (the vector operator), list (lsodes requires a list), and return. The final } concludes ode_1.

```
#
# Return derivative vector
  return(list(c(yt)));
  }
```

This completes the programming of eqs. (6.2). The numerical and graphical output is considered next.

(6.3) Model output

The numerical output from Listings 6.1 and 6.2 is in Table 6.1.

We can note the following details for this output.

- The ICs of eqs. (6.3) are confirmed. While this check may seem obvious, it is worth doing, since if the ICs are incorrect, the resulting solution will be incorrect.

t	y1(t)	y2(t)
0.0	1.0000	1.0000

- The toxin transfer from the extravascular volume (with solution y_1) to the intravascular volume (with solution y_2) is substantial. This transfer is in part due to the use of lam21=0.017 rather than lam21=0.0017 as originally specified in [1], p 397, Table 2.
- The final t, 360, and the spacing in t, 10, are as expected.

350.0	0.6851	0.5696
360.0	0.6847	0.5691

- The solution is smooth, e.g., nonnegative and nonoscillatory, as we would expect for the physical system.

```
VP = 5000.0   VE = 8333.3
lam12 =  0.00470  lam21 =  0.01700
QB = 150.0  QE =  0.01000  alpha =  1.00
y2in =  0.010  y10 =  1.000  y20 =  1.000
```

t	y1(t)	y2(t)
0.0	1.0000	1.0000
10.0	0.9608	0.9437
20.0	0.9264	0.8951
30.0	0.8961	0.8531
40.0	0.8696	0.8166
50.0	0.8464	0.7849
60.0	0.8260	0.7573
70.0	0.8081	0.7332
80.0	0.7925	0.7122
90.0	0.7788	0.6939
100.0	0.7668	0.6779
110.0	0.7562	0.6639
120.0	0.7470	0.6516
130.0	0.7390	0.6409
140.0	0.7319	0.6315
150.0	0.7257	0.6233
160.0	0.7203	0.6161
170.0	0.7155	0.6098
180.0	0.7114	0.6043
190.0	0.7077	0.5995
200.0	0.7045	0.5953
210.0	0.7017	0.5916
220.0	0.6993	0.5884
230.0	0.6971	0.5855
240.0	0.6953	0.5830
250.0	0.6936	0.5809
260.0	0.6922	0.5790
270.0	0.6909	0.5773
280.0	0.6898	0.5758
290.0	0.6889	0.5746
300.0	0.6880	0.5734
310.0	0.6873	0.5725
320.0	0.6866	0.5716
330.0	0.6861	0.5708
340.0	0.6856	0.5702
350.0	0.6851	0.5696
360.0	0.6847	0.5691

```
ncall =  65
```

Table 6.1: Numerical solution for eqs. (6.2) and (6.3)

- y_1 and y_2 approach steady state values of approximately 0.685 and 0.569 as also demonstrated in Fig. 6.2.
- The computational effort is quite modest as we might expect for a low-order ODE system with a smooth solution (also, `lsodes` is a stiff ODE integrator, just in case eqs. (6.2) are stiff).

```
ncall =   65
```

The plot in Fig. 6.2 confirms these conclusions.

To conclude this discussion of the ODE patient model, two terms in eqs. (6.2) move the solution through t: (1) the inhomogeneous term in eq. (6.2a), Q_E, models the ongoing production of toxin in the extravascular volume, and (2) the entering blood flow with concentration y_2^{in} in eq. (6.2b) represents the overall effect of the ALSS, that is, the reduction in the concentration y_2 to y_2^{in}. Q_E and y_2^{in}, defined as parameters in Listing 6.1, are therefore parameters of particular interest in studying the ODE patient model.

We now proceed to the addition of the two membrane units, MU1, MU2, and the two adsorption units, AU1, AU2, depicted in Fig. 6.3.

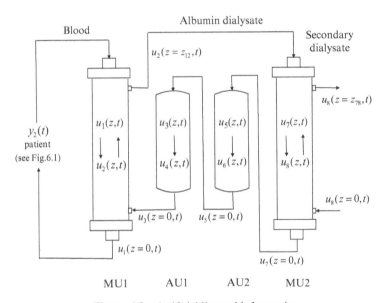

Figure 6.3 Artificial liver with four units

(6.4) 8-PDE ALSS model

The ALSS model is represented schematically in Fig. 6.3. The two membrane and two adsorption units have dependent variables (toxin concentrations) $u_1,...,u_8$.

(6.4.1) Membrane unit MU1

For MU1, the PDEs are[2]

$$\frac{\partial u_1}{\partial t} = -v_1 \frac{\partial u_1}{\partial z} + R_1 k_{12}(u_2 - u_1) \tag{6.4a}$$

$$\frac{\partial u_2}{\partial t} = -v_2 \frac{\partial u_2}{\partial z} + R_2 k_{12}(u_1 - u_2) \tag{6.4b}$$

where (Fig. 6.3)

u_1 interior membrane toxin concentration (in blood) (μmol/ml$_{fluid}$)
u_2 exterior membrane toxin concentration (in albumin dialysate) (μmol/ml$_{fluid}$)
z axial position along MU1 (cm$_{membrane}$)
t time (min)
v_1, v_2 velocities for u_1, u_2 (cm/min)
R_1 ratio of mass transfer area per unit length to flow cross sectional area for
             membrane interior
R_2 ratio of mass transfer area per unit length to flow cross sectional area for
             membrane exterior
k_{12} coefficient for mass transfer across the membrane (cm/min)

Axial dispersion is considered negligible so that eqs. (6.4a,b) do not have second-order derivatives in z.[3]

Eqs. (6.4a,b) are first order in z and t, and therefore require one BC and one IC.

$$u_1(z = z_{12}, t) = y_2(t); \ u_1(z, t = 0) = u_{10} \tag{6.4c,d}$$

$$u_2(z = 0, t) = u_3(z = 0, t); \ u_2(z, t = 0) = u_{20} \tag{6.4e,f}$$

BC (6.4c) specifies that the u_1 entering concentration (z_{12} is the axial length of MU1, Fig. 6.3) is the exiting patient concentration. BC (6.4e) specifies that the u_2 entering concentration is the exiting concentration of AU1. u_{10}, u_{20} are specified initial concentrations. Eqs. (6.4) are the PDE model for MU1.

(6.4.2) Adsorption unit AU1

For AU1, the PDEs are discussed as eqs. (A.6.7), (A.6.8) in the chapter appendix.

$$\frac{\partial u_3}{\partial t} = -v_3 \frac{\partial u_3}{\partial z} + D_3 \frac{\partial^2 u_3}{\partial z^2} - ((1 - \epsilon)/\epsilon)k_{34}(u_4^e - u_4) \tag{6.5a}$$

---

[2]Additional explanation for eq. (6.4a) as a mass balance on an incremental length of MU1 is given in the chapter appendix.
[3]Axial dispersion is discussed in the chapter appendix, and in particular, is included in eq. (A.6.5).

where

u_3	toxin concentration in albumin dialysate (μmol/ml$_{fluid}$)
u_4	toxin concentration on adsorbent (μmol/ml$_{adsorbent}$)
u_4^e	equilibrium absorbent concentration (μmol/ml$_{adsorbent}$)
z	axial position along AU1 (cm$_{adsorber}$)
t	time (min)
v_3	fluid velocity for u_3 (cm$_{adsorber}$/min)
D_3	dispersion coefficient for u_3 (cm$^2_{adsorber}$/min)
ϵ	absorbent void fraction (ml_{fluid}/$cm^3_{adsorber}$)
k_{34}	coefficient for mass transfer between fluid and adsorbent (min^{-1})

Note that in Fig. 6.3, the fluid flow in AU1 is in the negative direction (top to bottom) so that $v_3 < 0$.

Eq. (6.5a) is second order in z and t, and therefore requires two BCs and one IC.

$$u_3(z = z_{34}, t) = u_5(z = 0, t); \quad \frac{\partial u_3(z = 0, t)}{\partial t} = -v_3 \frac{\partial u_3(z = 0, t)}{\partial z} \quad \text{(6.5b,c)}$$

$$u_3(z, t = 0) = u_{30} \quad \text{(6.5d)}$$

BC (6.5b) specifies that the u_3 entering concentration is the exiting concentration of AU2 (z_{34} is the length of AU1). Eq. (6.5c) is a dynamic exiting BC[4] (dynamic in the sense that it has a derivative in t). u_{30} is a specified initial concentration.

For u_4, the PDE is

$$\frac{\partial u_4}{\partial t} = k_{34}(u_4^e - u_4) \quad \text{(6.5e)}$$

Eq. (6.5e) does not have derivatives in z (e.g., no flow or dispersion in the solid phase of AU1), but u_4 is still a function of z since it varies along the AU1 unit. Since eq. (6.5e) is first order in t, it requires one IC.

$$u_4(z, t = 0) = u_{40} \quad \text{(6.5f)}$$

This completes the model for AU1.

(6.4.3) Adsorption unit AU2

For AU2, the PDEs are analogous to eqs. (6.5).

$$\frac{\partial u_5}{\partial t} = -v_5 \frac{\partial u_5}{\partial z} + D_5 \frac{\partial^2 u_5}{\partial z^2} - ((1 - \epsilon)/\epsilon)k_{56}(u_6^e - u_6) \quad \text{(6.6a)}$$

---

[4]BC (6.5c) is used rather than the Danckwerts exiting BC that specifies zero slope which is physically impossible to realize. The use of dynamic BCs is discussed in [2].

where

u_5	toxin concentration in albumin dialysate (μmol/ml$_{fluid}$)
u_6	toxin concentration on adsorbent (μmol/ml$_{adsorbent}$)
u_6^e	equilibrium absorbent concentration (μmol/ml$_{adsorbent}$)
z	axial position along AU2 (cm$_{adsorber}$)
t	time (min)
v_5	fluid velocity for u_5 (cm$_{adsorber}$/min)
D_5	dispersion coefficient for u_3 (cm$^2_{adsorber}$/min)
ϵ	absorbent void fraction (ml$_{fluid}$/cm$^3_{adsorber}$)
k_{56}	coefficient for mass transfer between fluid and adsorbent (min^{-1})

Note that in Fig. 6.3, the fluid flow in AU2 is in the negative direction (top to bottom) so that $v_5 < 0$.

Eq. (6.6a) is second order in z and t, and therefore requires two BCs and one IC.

$$u_5(z = z_{56}, t) = u_7(z = 0, t); \frac{\partial u_5(z = 0, t)}{\partial t} = -v_5 \frac{\partial u_5(z = 0, t)}{\partial z} \qquad (6.6b,c)$$

$$u_5(z, t = 0) = u_{50} \qquad (6.6d)$$

BC (6.5b) specifies that the u_5 entering concentration is the exiting concentration of MU2 (z_{56} is the length of AU2). BC (6.6c) is a dynamic exiting BC [2]. u_{50} is a specified initial concentration.

For u_6, the PDE is

$$\frac{\partial u_6}{\partial t} = k_{34}(u_6^e - u_6) \qquad (6.6e)$$

Eq. (6.6e) is first order in t and requires one IC.

$$u_6(z, t = 0) = u_{60} \qquad (6.6f)$$

This completes the model for AU2.

(6.4.4) Membrane unit MU2

For MU2, the PDEs are analogous to those for MU1 (eqs. (6.4)).

$$\frac{\partial u_7}{\partial t} = -v_7 \frac{\partial u_7}{\partial z} + R_7 k_{78}(u_8 - u_7) \qquad (6.7a)$$

$$\frac{\partial u_8}{\partial t} = -v_8 \frac{\partial u_8}{\partial z} + R_8 k_{78}(u_7 - u_8) \qquad (6.7b)$$

where

u_7	interior membrane toxin concentration (in blood) (μmol/ml$_{fluid}$)
u_8	exterior membrane toxin concentration (in albumin dialysate) (μmol/ml$_{fluid}$)

z axial position along MU2 ($cm_{membrane}$)

t time (min)

v_7, v_8 velocities for u_7, u_8 (cm/min)

R_7 ratio of mass transfer area per unit length to flow cross sectional area for membrane interior

R_8 ratio of mass transfer area per unit length to flow cross sectional area for membrane exterior

k_{78} coefficient for mass transfer across the membrane (cm/min)

Axial dispersion is considered negligible so that eqs. (6.7) do not have second-order derivatives in z

Eqs. (6.7a,b) are first order in z and t, and therefore require one BC and one IC.

$$u_7(z = z_{78}, t) = u_2(z = z_{12}, t); \ u_7(z, t = 0) = u_{70} \qquad (6.7\text{c,d})$$

$$u_8(z = 0, t) = u_{8e}(t); \ u_8(z, t = 0) = u_{80} \qquad (6.7\text{e,f})$$

BC (6.7c) specifies that the u_7 entering concentration (z_{78} is the axial length of MU2, Fig. 6.3) is the exiting concentration from MU1. BC (6.7e) specifies that the u_8 entering concentration is a function of t, $u_{8e}(t)$. u_{70}, u_{80} are specified initial concentrations. Eqs. (6.7) are the PDE model for MU2.

(6.5) Patient-ALSS ODE/PDE model routines

The addition of eqs. (6.4)-(6.7) to eqs. (6.2), (6.3) constitutes the 2-ODE 8-PDE patient-ALSS model. The main program is in Listing 6.3.

(6.5.1) Main program

```
#
# 2-ODE patient, 8-PDE artificial liver support
# system model
#
# Delete previous workspaces
  rm(list=ls(all=TRUE))
#
# Access ODE integrator
  library("deSolve");
#
# Access functions for numerical solution
  setwd("g:/chap6");
  source("pde_1.R");
  source("dss020.R");
  source("dss044.R");
#
```

```
# Level of output
#
#   ip = 1 - graphical (plotted) solutions
#             (u1(x,t), u2(x,t)) only
#
#   ip = 2 - numerical and graphical solutions
#
  ip=2;
#
# Parameters
#
# 2-ode model
    VP=5000; VE=(5/3)*VP; lam12=0.0047; lam21=0.0017; QB=150;
   QE=0.005;       alpha=1;    y2in=0.01;          y10=1; y20=1;
  lam21=0.017;
  cat(sprintf("\n  VP = %4.1f   VE = %4.1f\n",VP,VE));
  cat(sprintf("\n  lam12 = %8.5f  lam21 = %8.5f\n",lam12,lam21));
  cat(sprintf("\n  QB = %4.1f  QE = %8.5f  alpha = %5.2f\n",
              QB,QE,alpha));
  cat(sprintf("\n  y2in = %5.3f  y10 = %5.3f  y20 = %5.3f\n",
              y2in,y10,y20));
#
# 8-pde model
#
# Initial concentrations
  u10=0;u20=0;u30=0;u40=0;
  u50=0;u60=0;u70=0;u80=0;
#
# u1,u2
  z12=25;v1=-2.5;v2=2.5;
  R1=1;R2=1;k12=0.05;
#
# u3,u4
  z34=25;eps=0.3;k1=1;k2=10;
  v3=-2.5;D3=0.01;k34=0.05;
#
# u5,u6
  z56=25;
  v5=-2.5;D5=0.01;k56=0.05;
#
# u7,u8
  z78=25;v7=-2.5;v8=2.5;u8e=0;
  R7=1;R8=1;k78=0.05;
#
# Independent variable for ODE integration
  nout=37;t0=0;tf=360;
```

```
    tout=seq(from=t0,to=tf,by=(tf-t0)/(nout-1));
#
# Initial conditions
  nz=21;
  u0=rep(0,2+8*nz);
  for(i in 1:nz){
    u0[i]=u10;
    u0[i+nz]=u20;
    u0[i+2*nz]=u30;
    u0[i+3*nz]=u40;
    u0[i+4*nz]=u50;
    u0[i+5*nz]=u60;
    u0[i+6*nz]=u70;
    u0[i+7*nz]=u80;
  }
  u0[1+8*nz]=y10;
  u0[2+8*nz]=y20;
  ncall=0;
#
# ODE integration
  out=lsodes(func=pde_1,y=u0,times=tout,parms=NULL)
#
# Save and display numerical solution
  u1=rep(0,nout);u2=rep(0,nout);u3=rep(0,nout);u4=rep(0,nout);
  u5=rep(0,nout);u6=rep(0,nout);u7=rep(0,nout);u8=rep(0,nout);
  y1=rep(0,nout);y2=rep(0,nout);
  for(it in 1:nout){
    u1[it]=out[it,2]      ;u2[it]=out[it,2*nz+1];
    u3[it]=out[it,2*nz+2];u5[it]=out[it,4*nz+2];
    u4[it]=out[it,3*nz+2];u6[it]=out[it,5*nz+2];
    u7[it]=out[it,6*nz+2];u8[it]=out[it,8*nz+1];
    y1[it]=out[it,8*nz+2];y2[it]=out[it,8*nz+3];
  }
  if(ip==2){
    for(it in 1:nout){
      if(it==1){
        cat(sprintf("\n     t      y1(t)      y2(t)\n"));
      }
      cat(sprintf("%7.1f%10.4f%10.4f\n",tout[it],y1[it],y2[it]));
    }
  }
#
# Calls to ODE routine
  cat(sprintf("\n\n  ncall = %3d\n\n",ncall));
#
# Plot y1, y2
```

```
  par(mfrow=c(1,1))
  plot(tout,y1,
     xlab="t (min)",ylab="y1,y2 (micromol/ml)",
     main="y1(t),y2(t)",type="l",lwd=2,
     xlim=c(0,400),ylim=c(0.4,1.3));
   points(tout,y1, pch="1",lwd=2);
    lines(tout,y2,type="l",lwd=2);
    points(tout,y2, pch="2",lwd=2);
#
# Plot u1(z=0,t), u2(z=z12,t)
  par(mfrow=c(1,1))
  plot(tout,u1,
     xlab="t (min)",ylab="u1,u2 (micromol/ml)",
     main="u1(z=0,t),u2(z=z12,t)",type="l",lwd=2,
     xlim=c(0,400));
   points(tout,u1, pch="1",lwd=2);
    lines(tout,u2,type="l",lwd=2);
    points(tout,u2, pch="2",lwd=2);
#
# Plot u3(z=0,t), u5(z=0,t)
  par(mfrow=c(1,1))
  plot(tout,u3,
     xlab="t (min)",ylab="u3,u5 (micromol/ml)",
     main="u3(z=0,t),u5(z=0,t)",type="l",lwd=2,
     xlim=c(0,400));
   points(tout,u3, pch="3",lwd=2);
    lines(tout,u5,type="l",lwd=2);
    points(tout,u5, pch="5",lwd=2);
#
# Plot u4(z=0,t), u6(z=0,t)
  par(mfrow=c(1,1))
  plot(tout,u4,
     xlab="t (min)",ylab="u4,u6 (micromol/ml)",
     main="u4(z=0,t),u6(z=0,t)",type="l",lwd=2,
     xlim=c(0,400));
   points(tout,u4, pch="4",lwd=2);
    lines(tout,u6,type="l",lwd=2);
    points(tout,u6, pch="6",lwd=2);
#
# Plot u7(z=0,t), u8(z=z78,t)
  par(mfrow=c(1,1))
  plot(tout,u7,
     xlab="t (min)",ylab="u7,u8 (micromol/ml)",
     main="u7(z=0,t),u8(z=z78,t)",type="l",lwd=2,
     xlim=c(0,400));
   points(tout,u7, pch="7",lwd=2);
```

```
lines(tout,u8,type="l",lwd=2);
points(tout,u8, pch="8",lwd=2);
```

Listing 6.3: Main program for eqs. (6.2) and (6.3)

Listing 6.3 is similar to Listing 6.1 so only the differences will be considered. We can note the following details.

- Previous files are removed

```
#
# 2-ODE patient, 8-PDE artificial liver support
# system model
#
# Delete previous workspaces
  rm(list=ls(all=TRUE))
```

- The files accessed by source include the ODE/MOL routine, pde_1, and two differentiation in space (DSS) routines, dss020, dss044.

```
#
# Access functions for numerical solution
  setwd("g:/chap6");
  source("pde_1.R");
  source("dss020.R");
  source("dss044.R");
```

- The ODE parameters are the same as in Listing 6.1. Again, the value of lam21 is changed from 0.0017 to 0.017 to give a more pronounced transfer of toxin between the extravascular (tissue) and intravascular (blood) volumes.

```
#
# Parameters
#
# 2-ode model
    VP=5000; VE=(5/3)*VP; lam12=0.0047; lam21=0.0017; QB=150;
  QE=0.005;      alpha=1;     y2in=0.01;          y10=1;  y20=1;
  lam21=0.017;
  cat(sprintf("\n  VP = %4.1f   VE = %4.1f\n",VP,VE));
  cat(sprintf("\n  lam12 = %8.5f  lam21 = %8.5f\n",lam12,
          lam21));
  cat(sprintf("\n  QB = %4.1f   QE = %8.5f   alpha = %5.2f\n",
              QB,QE,alpha));
  cat(sprintf("\n  y2in = %5.3f  y10 = %5.3f  y20 = %5.3f\n",
              y2in,y10,y20));
```

- The PDE ICs and parameters for eqs. (6.4) to (6.7) are then added.

```
#
# 8-PDE model
#
# Initial concentrations
  u10=0;u20=0;u30=0;u40=0;
  u50=0;u60=0;u70=0;u80=0;
#
# u1,u2
  z12=25;v1=-2.5;v2=2.5;
  R1=1;R2=1;k12=0.05;
#
# u3,u4
  z34=25;eps=0.3;k1=1;k2=10;
  v3=-2.5;D3=0.01;k34=0.05;
#
# u5,u6
  z56=25;
  v5=-2.5;D5=0.01;k56=0.05;
#
# u7,u8
  z78=25;v7=-2.5;v8=2.5;u8e=0;
  R7=1;R8=1;k78=0.05;
```

- After definition of the t scale as in Listing 6.1 (vector tout), an IC vector, u0, of length 2+8*nz is defined. nz=21 defines the number of grid points in z for each of eqs. (6.4) to (6.7). The length of MU1, AU1, AU2, MU2 in Fig. 6.3 is 25 cm. The grid in z therefore has a spacing $25/(21 - 1) = 1.25$ so that $z = 0, 1.25,...,25$ in the MOL approximation.

  The velocities v_1, v_3, v_5, v_7 are negative, and the velocities v_2, v_8 are positive as indicated in Fig. 6.3. AU1, AU2 have only one flowing stream (with velocities v_3, v_5) since the adsorbent phase is immobile. The mass transfer coefficients are $k_{12} = k_{34} = k_{56} = k_{78} = 0.05$.

  ICs (6.4d,f), (6.5d,f), (6.6d,f) and (6.7d,f) are defined through the initial values u10, . . . , u80. ICs (6.3a,b) are then defined.

```
#
# Initial conditions
  nz=21;
  u0=rep(0,2+8*nz);
  for(i in 1:nz){
    u0[i]=u10;
    u0[i+nz]=u20;
    u0[i+2*nz]=u30;
    u0[i+3*nz]=u40;
```

```
    u0[i+4*nz]=u50;
    u0[i+5*nz]=u60;
    u0[i+6*nz]=u70;
    u0[i+7*nz]=u80;
  }
  u0[1+8*nz]=y10;
  u0[2+8*nz]=y20;
  ncall=0;
```

After the 8*nz PDE dependent variable values for eqs. (6.4) to (6.7), y_1, y_2 from eqs. (6.2) are placed in u0 as elements 1+8*nz,2+8*nz. Thus, the total number of ODEs is 2+8*nz = 170. This length is used to tell lsodes the total number of ODEs to be integrated.

- lsodes integrates the 170-ODE system. The MOL/ODE routine is pde_1 that is discussed subsequently. The input parameters are discussed for Listing 6.1.

```
#
# ODE integration
  out=lsodes(func=pde_1,y=u0,times=tout,parms=NULL)
```

The numerical solution of the 170-ODE system is returned in out.

- Selected PDE dependent variables of eqs. (6.4) to (6.7) are placed in arrays u1 to u8 for plotting. Generally, these are the exiting or outflow stream concentrations in Fig. 6.3. For example, $u_1(z = 0, t) = $ out[it,2] is from the bottom of MU1. The offset of 1 in the second subscript such as [it,2] is required since out[it,1] is reserved for the independent variable t. As another example, $u_2(z = z_{12}, t) = $ out[it,2*nz+1] is from the top of MU1. The ODE dependent variables of eqs. (6.2) are placed in arrays y1,y2. All of these arrays are first declared with the rep utility.

```
#
# Save and display numerical solution
  u1=rep(0,nout);u2=rep(0,nout);
  u3=rep(0,nout);u4=rep(0,nout);
  u5=rep(0,nout);u6=rep(0,nout);
  u7=rep(0,nout);u8=rep(0,nout);
  y1=rep(0,nout);y2=rep(0,nout);
  for(it in 1:nout){
    u1[it]=out[it,2]       ;u2[it]=out[it,2*nz+1];
    u3[it]=out[it,2*nz+2];u5[it]=out[it,4*nz+2];
    u4[it]=out[it,3*nz+2];u6[it]=out[it,5*nz+2];
    u7[it]=out[it,6*nz+2];u8[it]=out[it,8*nz+1];
    y1[it]=out[it,8*nz+2];y2[it]=out[it,8*nz+3];
  }
  if(ip==2){
    for(it in 1:nout){
      if(it==1){
```

```
            cat(sprintf("\n       t      y1(t)      y2(t)\n"));
          }
        cat(sprintf("%7.1f%10.4f%10.4f\n",tout[it],y1[it],
                y2[it]));
      }
    }
```

For ip=2 the solutions to eqs. (6.2) are displayed numerically as before in Listing 6.1.

- The two ODE dependent variables in arrays y1,y2 and eight PDE variables in u1,...,u8 are plotted against t with the plot,points,lines utilities as explained for Listing 6.1. In each plot, two curves are plotted and identified with pch.

```
#
# Plot y1, y2
  par(mfrow=c(1,1))
  plot(tout,y1,
      xlab="t (min)",ylab="y1,y2 (micromol/ml)",
      main="y1(t),y2(t)",type="l",lwd=2,
      xlim=c(0,400),ylim=c(0.4,1.3));
    points(tout,y1, pch="1",lwd=2);
     lines(tout,y2,type="l",lwd=2);
    points(tout,y2, pch="2",lwd=2);
#
# Plot u1(z=0,t), u2(z=z12,t)
  par(mfrow=c(1,1))
  plot(tout,u1,
      xlab="t (min)",ylab="u1,u2 (micromol/ml)",
      main="u1(z=0,t),u2(z=z12,t)",type="l",lwd=2,
      xlim=c(0,400));
    points(tout,u1, pch="1",lwd=2);
     lines(tout,u2,type="l",lwd=2);
    points(tout,u2, pch="2",lwd=2);

             .              .
             .              .
             .              .

#
# Plot u7(z=0,t), u8(z=z78,t)
  par(mfrow=c(1,1))
  plot(tout,u7,
      xlab="t (min)",ylab="u7,u8 (micromol/ml)",
      main="u7(z=0,t),u8(z=z78,t)",type="l",lwd=2,
      xlim=c(0,400));
    points(tout,u7, pch="7",lwd=2);
     lines(tout,u8,type="l",lwd=2);
    points(tout,u8, pch="8",lwd=2);
```

This completes the programming of the main program. The MOL/ODE routine pde_1 called by lsodes is discussed next.

(6.5.2) ODE routine

MOL/ODE routine pde_1 follows.

```
  pde_1=function(t,u,parms){
#
# Function pde_1 computes the t derivative vector of
# u1(t),..., u8(t), y1(t), y2(t)
#
# One vector to eight vectors and two scalars
  u1=rep(0,nz);u2=rep(0,nz);
  u3=rep(0,nz);u4=rep(0,nz);
  u5=rep(0,nz);u6=rep(0,nz);
  u7=rep(0,nz);u8=rep(0,nz);
  for(i in 1:nz){
    u1[i]=u[i]       ;u2[i]=u[i+nz]   ;
    u3[i]=u[i+2*nz];u4[i]=u[i+3*nz];
    u5[i]=u[i+4*nz];u6[i]=u[i+5*nz];
    u7[i]=u[i+6*nz];u8[i]=u[i+7*nz];
  }
  y1=u[1+8*nz];y2=u[2+8*nz];
#
# PDEs
#
# u1t, u2t
  u1t=rep(0,nz);u2t=rep(0,nz);
  u1[nz]=y2;u2[1]=u3[nz];
  u1z=dss020(0,z12,nz,u1,v1);
  u2z=dss020(0,z12,nz,u2,v2);
  for(i in 1:nz){
    u1t[i]=-v1*u1z[i]+R1*k12*(u2[i]-u1[i]);
    u2t[i]=-v2*u2z[i]+R2*k12*(u1[i]-u2[i]);
  }
  u1t[nz]=0;u2t[1]=0;
#
# u3t, u4t
  u3t=rep(0,nz);u4t=rep(0,nz);
  u3z=dss020(0,z34,nz,u3,v3);
  nl=1;nu=1;
  u3zz=dss044(0,z34,nz,u3,u3z,nl,nu);
  epsr=(1-eps)/eps;
  for(i in 2:(nz-1)){
    u4e=k1*u3[i]/(1+k2*u3[i]);
```

```
      u3t[i]=-v3*u3z[i]+D3*u3zz[i]-epsr*k34*(u4e-u4[i]);
      u4t[i]=k34*(u4e-u4[i]);
    }
  u3[nz]=u5[1];u3t[nz]=0;
  u3t[1]=-v3*u3z[1];
  u4e=k1*u3[1]/(1+k2*u3[1]);
  u4t[1]=k34*(u4e-u4[1]);
  u4e=k1*u3[nz]/(1+k2*u3[nz]);
  u4t[nz]=k34*(u4e-u4[nz]);
#
# u5t, u6t
  u5t=rep(0,nz);u6t=rep(0,nz);
  u5z=dss020(0,z56,nz,u5,v5);
  nl=1;nu=1;
  u5zz=dss044(0,z56,nz,u5,u5z,nl,nu);
  for(i in 2:(nz-1)){
    u6e=k1*u6[i]/(1+k2*u6[i]);
    u5t[i]=-v5*u5z[i]+D5*u5zz[i]-epsr*k56*(u6e-u6[i]);
    u6t[i]=k56*(u6e-u6[i]);
    }
  u5[nz]=u7[1];u5t[nz]=0;
  u5t[1]=-v5*u5z[1];
  u6e=k1*u5[1]/(1+k2*u5[1]);
  u6t[1]=k56*(u6e-u6[1]);
  u6e=k1*u5[nz]/(1+k2*u5[nz]);
  u6t[nz]=k56*(u6e-u6[nz]);
#
# u7t, u8t
  u7t=rep(0,nz);u8t=rep(0,nz);
  u7[nz]=u2[nz];u8[1]=u8e;
  u7z=dss020(0,z78,nz,u7,v7);
  u8z=dss020(0,z78,nz,u8,v8);
  for(i in 1:nz){
    u7t[i]=-v7*u7z[i]+R7*k78*(u8[i]-u7[i]);
    u8t[i]=-v8*u8z[i]+R8*k78*(u7[i]-u8[i]);
    }
  u7t[nz]=0;u8t[1]=0;
#
# ODEs
# y1, y2
  y2u=alpha*y2;
  y1t=(VP/VE)*lam12*y2u-lam21*y1+QE;
  y2t=(QB/VP)*(u1[1]-y2)-lam12*y2u+(VE/VP)*lam21*y1;
#
# Eight vectors and two scalars to one vector
  ut=rep(0,2+8*nz);
```

```
  for(i in 1:nz){
    ut[i]=u1t[i]        ;ut[i+nz]=u2t[i]   ;
    ut[i+2*nz]=u3t[i];ut[i+3*nz]=u4t[i];
    ut[i+4*nz]=u5t[i];ut[i+5*nz]=u6t[i];
    ut[i+6*nz]=u7t[i];ut[i+7*nz]=u8t[i];
  }
  ut[1+8*nz]=y1t;ut[2+8*nz]=y2t;
#
# Increment calls to pde_1
  ncall <<- ncall+1;
#
# Return derivative vector
  return(list(c(ut)));
  }
```

Listing 6.4: MOL/ODE routine for eqs. (6.2) to (6.7)

We can note the following details about Listing 6.4.

- The function is defined.

```
    pde_1=function(t,u,parms){
  #
  # Function pde_1 computes the t derivative vector of
  # u1(t),..., u8(t), y1(t), y2(t)
```

  u is a 170-vector of ODE dependent variables. t is the current value of t in eqs. (6.2) to (6.7). parms for passing parameters to pde_1 is unused.

- u is placed in eight vectors, u1, . . . , u8 of length nz, and two scalars y1, y2. The PDE vectors are first declared (preallocated) with the rep utility.

```
    #
    # One vector to eight vectors and two scalars
      u1=rep(0,nz);u2=rep(0,nz);
      u3=rep(0,nz);u4=rep(0,nz);
      u5=rep(0,nz);u6=rep(0,nz);
      u7=rep(0,nz);u8=rep(0,nz);
      for(i in 1:nz){
        u1[i]=u[i]        ;u2[i]=u[i+nz]   ;
        u3[i]=u[i+2*nz];u4[i]=u[i+3*nz];
        u5[i]=u[i+4*nz];u6[i]=u[i+5*nz];
        u7[i]=u[i+6*nz];u8[i]=u[i+7*nz];
      }
      y1=u[1+8*nz];y2=u[2+8*nz];
```

- The PDEs, eqs. (6.4) to (6.7), are programmed, starting with eqs. (6.4).

```
#
# PDEs
#
# u1t, u2t
  u1t=rep(0,nz);u2t=rep(0,nz);
  u1[nz]=y2;u2[1]=u3[nz];
  u1z=dss020(0,z12,nz,u1,v1);
  u2z=dss020(0,z12,nz,u2,v2);
  for(i in 1:nz){
    u1t[i]=-v1*u1z[i]+R1*k12*(u2[i]-u1[i]);
    u2t[i]=-v2*u2z[i]+R2*k12*(u1[i]-u2[i]);
  }
  u1t[nz]=0;u2t[1]=0;
```

We can note the following details about this programming.

- Arrays u1t,u2t are declared for the partial derivatives $\partial u_1/\partial t, \partial u_2/\partial t$ in eqs. (6.4a,b).

    ```
    u1t=rep(0,nz);u2t=rep(0,nz);
    ```

- BCs (6.4c,e) are programmed as (refer to Fig. 6.3)

    ```
    u1[nz]=y1;u2[1]=u3[nz];
    ```

- Partial derivatives $\partial u_1/\partial z, \partial u_2/\partial z$ in eqs. (6.4a,b) are computed by dss020 (arrays u1z,u2z, which do not have to be declared since this is done in dss020).

    ```
    u1z=dss020(0,z12,nz,u1,v1);
    u2z=dss020(0,z12,nz,u2,v2);
    ```

    The velocities v1,v2 are used by dss020 to define the direction of flow (the direction of biasing or upwinding of the five-point finite differences in dss020). That is, v1<0 since the flow for u_1 is from $z = z_{12}$ to $z = 0$ while v2>0 since the flow for u_2 is from $z = 0$ to $z = z_{12}$. Incorrect specification of the direction of flow (sign of the velocity) generally leads to an unstable numerical solution.

- Eqs. (6.4a,b) are programmed.

    ```
    for(i in 1:nz){
      u1t[i]=-v1*u1z[i]+R1*k12*(u2[i]-u1[i]);
      u2t[i]=-v2*u2z[i]+R2*k12*(u1[i]-u2[i]);
    }
    u1t[nz]=0;u2t[1]=0;
    ```

    Since $u_1(z = z_{12}, t) = y_2$, $u_2(z = 0, t) = u_3(z = 0, t)$, are set by BCs, the derivatives in t are set to zero. The close correspondence of the programming and the PDEs (eqs. (6.4a,b)) demonstrates an important advantage of the MOL.

- Eqs. (6.5) are programmed in a similar way.

```
#
# u3t, u4t
  u3t=rep(0,nz);u4t=rep(0,nz);
  u3z=dss020(0,z34,nz,u3,v3);
  nl=1;nu=1;
  u3zz=dss044(0,z34,nz,u3,u3z,nl,nu);
  epsr=(1-eps)/eps;
  for(i in 2:(nz-1)){
    u4e=k1*u3[i]/(1+k2*u3[i]);
    u3t[i]=-v3*u3z[i]+D3*u3zz[i]-epsr*k34*(u4e-u4[i]);
    u4t[i]=k34*(u4e-u4[i]);
  }
  u3[nz]=u5[1];u3t[nz]=0;
  u3t[1]=-v3*u3z[1];
  u4e=k1*u3[1]/(1+k2*u3[1]);
  u4t[1]=k34*(u4e-u4[1]);
  u4e=k1*u3[nz]/(1+k2*u3[nz]);
  u4t[nz]=k34*(u4e-u4[nz]);
```

We can note the following details.

– Vectors u3t, u4t are for the derivatives $\partial u_3/\partial t$, $\partial u_4/\partial t$ in eqs. (6.5a,b). $\partial u_3/\partial z$ in eq. (6.5a) is then computed by dss020 with v3<0 (for flow from $z = z_{34}$ to $z = 0$).

```
#
# u3t, u4t
  u3t=rep(0,nz);u4t=rep(0,nz);
  u3z=dss020(0,z34,nz,u3,v3);
```

– $\partial^2 u_3/\partial z^2$ in eq. (6.5a) is computed with dss044. nl=1, nu=1 specifies that the dependent variable u_3 at the boundaries $z = 0, z_{34}$ is used in calculating the second derivative (a *Dirichlet* BC). nl=2, nu=2 would specify *Neumann* BCs that use $\partial u_3/\partial z$ at the boundaries.

```
  nl=1;nu=1;
  u3zz=dss044(0,z34,nz,u3,u3z,nl,nu);
```

– Eqs. (6.5a,e) are programmed as u3t, u4t for $\partial u_3/\partial t$, $\partial u_4/\partial t$. First, the equilibrium concentration u_4^e in eqs. (6.5a,e) is computed from u_3 by a *Langmuir* isotherm (with parameters k_1, k_2).

```
  epsr=(1-eps)/eps;
  for(i in 2:(nz-1)){
    u4e=k1*u3[i]/(1+k2*u3[i]);
```

```
            u3t[i]=-v3*u3z[i]+D3*u3zz[i]-epsr*k34*(u4e-u4[i]);
            u4t[i]=k34*(u4e-u4[i]);
        }
```

Note that with for(i in 2:(nz-1)){, the boundary derivatives in t are not computed so far.

- $u_3(z = z_{34}, t)$ is set according to BC (6.5b), and its derivative, $\partial u_3(z = z_{34}, t)/\partial t$, is set to zero so the ODE integrator does not move it away from the specified value.

```
        u3[nz]=u5[1];u3t[nz]=0;
```

- $\partial u_3(z = 0, t)/\partial t$ is set according to BC (6.5c).

```
        u3t[1]=-v3*u3z[1];
```

u3z[1] was set by the preceding call to dss020.

- $\partial u_4(z = 0, t)/\partial t$ is set by eq. (6.5e), including the use of the isotherm for the equilibrium concentration u_4^e corresponding to $u_3(z = 0, t)$.

```
        u4e=k1*u3[1]/(1+k2*u3[1]);
        u4t[1]=k34*(u4e-u4[1]);
```

- $\partial u_4(z = z_{34}, t)/\partial t$ is set by eq. (6.5e), including the use of the isotherm for the equilibrium concentration u_4^e corresponding to $u_3(z = z_{34}, t)$.

```
        u4e=k1*u3[nz]/(1+k2*u3[nz]);
        u4t[nz]=k34*(u4e-u4[nz]);
```

- The programming of eqs. (6.6a,e) for $\partial u_5/\partial t$, $\partial u_6/\partial t$ is similar, including BCs (6.6b,c).

```
#
# u5t, u6t
  u5t=rep(0,nz);u6t=rep(0,nz);
        .           .
        .           .
        .           .
  u6t[nz]=k56*(u6e-u6[nz]);
```

- The programming of eqs. (6.7a,b) for $\partial u_7/\partial t$, $\partial u_8/\partial t$ is similar to the programming for $\partial u_1/\partial t$, $\partial u_2/\partial t$, including BCs (6.7c,e).

```
#
# u7t, u8t
  u7t=rep(0,nz);u8t=rep(0,nz);
```

```
u7[nz]=u2[nz];u8[1]=u8e;
u7z=dss020(0,z78,nz,u7,v7);
u8z=dss020(0,z78,nz,u8,v8);
for(i in 1:nz){
  u7t[i]=-v7*u7z[i]+R7*k78*(u8[i]-u7[i]);
  u8t[i]=-v8*u8z[i]+R8*k78*(u7[i]-u8[i]);
}
u7t[nz]=0;u8t[1]=0;
```

- Eqs. (6.2) are programmed as in Listing 6.2.

```
#
# ODEs
# y1, y2
  y2u=alpha*y2;
  y1t=(VP/VE)*lam12*y2u-lam21*y1+QE;
  y2t=(QB/VP)*(u1[1]-y2)-lam12*y2u+(VE/VP)*lam21*y1;
```

This completes the programming of the set of 170 ODE derivatives in t.

- The derivatives $\partial u_1/\partial t,..., \partial u_8/\partial t$ are placed in a derivative vector ut to be returned to the ODE integrator (lsodes called by the main program in Listing 6.3).

```
#
# Eight vectors and two scalars to one vector
  ut=rep(0,2+8*nz);
  for(i in 1:nz){
    ut[i]=u1t[i]       ;ut[i+nz]=u2t[i]   ;
    ut[i+2*nz]=u3t[i];ut[i+3*nz]=u4t[i];
    ut[i+4*nz]=u5t[i];ut[i+5*nz]=u6t[i];
    ut[i+6*nz]=u7t[i];ut[i+7*nz]=u8t[i];
  }
```

- The derivatives dy_1/dt, dy_2/dt (from eqs. (6.2)) are placed at the end of ut.

```
  ut[1+8*nz]=y1t;ut[2+8*nz]=y2t;
```

- The counter for the calls to pde_1 is returned to the main program of Listing 6.3 by a <<-.

```
#
# Increment calls to pde_1
  ncall <<- ncall+1;
```

- The derivative vector ut is returned to lsodes by a combination of c, the R vector operator, list since lsodes requires a list, and return.

```
#
# Return derivative vector
  return(list(c(ut)));
  }
```

The final } concludes pde_1. In summary, pde_1 receives a vector of 170 ODE dependent variables, u, and returns a vector of 170 derivatives, ut. Further, the dependent variables and their derivatives must be placed in corresponding positions in u and ut. That is, dependent variable u[i], with IC u0[i], has the associated derivative ut[i], where the index i has as a value in the interval i=1 to i=170.

Intermediate variables can be computed from the incoming u (such as u4e calculated from u3), then used in the calculation of ut. The intermediate variables are generally algebraic so that in this sense we are considering a *differential-algebraic equation (DAE)* system.[5]

This completes the programming of eqs. (6.2) to (6.7). The output from the preceding routines in Listings 6.3 and 6.4 is considered next.

(6.6) Model output

The numerical output from the main program of Listing 6.3 follows.

We can note the following details about this output.

- The ICs for y_1, y_2 are confirmed (at $t = 0$).
- The solutions for y_1, y_2 differ significantly from those in Table 6.1, which is due to the addition of MU1, AU1, AU2, MU2 to the 2-ODE patient model.
- The added units remove a significant part of the toxin in y_1 (tissue). For example, $y_1(t = 0) = 1.0000$ while $y_1(t = 360) = 0.4548$. y_2 (blood) goes through a maximum, then also drops below its IC, $y_2(t = 0) = 1.0000$, $y_2(t = 360) = 0.9196$.
- The computational effort with ncall = 796 is modest.

In summary, the ALSS is giving the expected response, i.e., a reduction in the toxin. This results primarily from the term in eq. (6.2b) that gives the patient-ALSS linkage. Specifically, with just y_1, y_2, the model is responding in eq. (6.2b) to $(y_2^{in} - y_2)$ with y2in = 0.01 in Listing 6.1. With MU1, AU1, AU2, MU2 added, the model is responding (in Listing 6.4) to (u1[1]-y2) so that the output of MU1 (u1[1]) becomes the input to the patient model. Therefore, a plot of $u_1(z = 0, t)$ is of interest and is discussed with the graphical output in Figs. 6.4a,b,c,d,e (refer also to Fig. 6.3).

Fig. 6.4a reflects the numerical solution of Table 6.2. In particular, the reduction in tissue toxin, $y_1(t)$, is clear.

---

[5] An extensive literature for DAE systems is available that pertains to a generalization of this basic idea, including many additional mathematical concepts. The discussion in this book is limited to cases in which the intermediate algebraic variables can be calculated explicitly in the MOL/ODE routine, as in pde_1 of Listing 6.4.

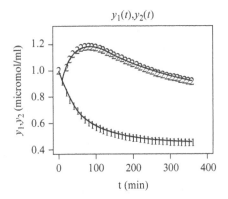

Figure 6.4a $y_1(t), y_2(t)$ from eqs. (6.2) and (6.3)

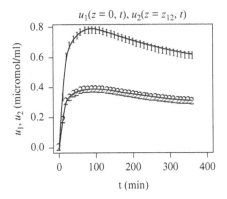

Figure 6.4b $u_1(z = 0, t), u_2(z = z_{12}, t)$ from eqs. (6.4)

Fig. 6.4b indicates the toxin concentration leaving the bottom of MU1 that becomes the input to the patient model, $u_1(z = 0, t)$, as discussed above. $u_2(z = z_{12}, t)$ indicates the uptake of toxin by MU1.

Fig. 6.4c indicates that the exiting streams from AU1 and AU2 have essentially zero toxin concentration. This is a desirable response since the adsorbent in AU1 and AU2 apparently does not have a level of toxin that would require regeneration.

Fig. 6.4d confirms the previous conclusion for Fig. 6.4c since the adsorbent at the bottom (exit) of AU1 and AU2 has essentially no toxin.

Fig. 6.4e indicates that the exiting stream from MU2 with concentration $u_7(z = 0, t)$ goes through a maximum of about 0.33, then declines as the transient from the IC $u_7(z, t = 0)$ progresses. Fig. 6.4c indicates that $u_7(z = 0, t)$ is essentially reduced to zero in AU2. The exiting dialysate stream from MU2, $u_8(z = z_{78}, t)$, approaches a steady state concentration of approximately 0.12.

(6.7) Summary and conclusions

In summary, the interactions of the patient, MU1, AU1, AU2 and MU2 models of eqs. (6.2) to (6.7) can be studied in detail by a MOL analysis. Once the composite model is

```
VP = 5000.0   VE = 8333.3
lam12 =  0.00470  lam21 =  0.01700
QB = 150.0   QE =  0.00500  alpha =   1.00
y2in = 0.010   y10 = 1.000   y20 = 1.000
```

t	y1(t)	y2(t)
0.0	1.0000	1.0000
10.0	0.9147	0.9377
20.0	0.8433	1.0264
30.0	0.7849	1.0852
40.0	0.7369	1.1269
50.0	0.6973	1.1546
60.0	0.6644	1.1715
70.0	0.6371	1.1801
80.0	0.6141	1.1823
90.0	0.5947	1.1798
100.0	0.5782	1.1738
110.0	0.5641	1.1652
120.0	0.5520	1.1547
130.0	0.5415	1.1430
140.0	0.5323	1.1306
150.0	0.5242	1.1176
160.0	0.5170	1.1045
170.0	0.5107	1.0915
180.0	0.5049	1.0786
190.0	0.4998	1.0660
200.0	0.4951	1.0537
210.0	0.4909	1.0419
220.0	0.4870	1.0306
230.0	0.4834	1.0197
240.0	0.4801	1.0093
250.0	0.4771	0.9994
260.0	0.4743	0.9900
270.0	0.4717	0.9811
280.0	0.4692	0.9726
290.0	0.4670	0.9646
300.0	0.4649	0.9570
310.0	0.4629	0.9499
320.0	0.4610	0.9431
330.0	0.4593	0.9367
340.0	0.4577	0.9307
350.0	0.4562	0.9250
360.0	0.4548	0.9196

```
ncall = 796
```

Table 6.2: Numerical solution for eqs. (6.2) and (6.3) with eqs. (6.4) to (6.7) included

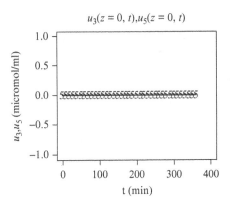

Figure 6.4c $u_3(z = 0, t), u_5(z = 0, t)$ from eqs. (6.5), (6.6)

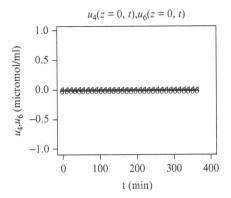

Figure 6.4d $u_4(z = 0, t), u_6(z = 0, t)$ from eqs. (6.5), (6.6)

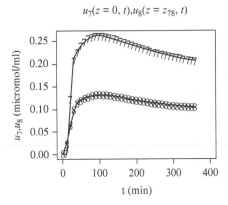

Figure 6.4e $u_7(z = 0, t), u_8(z = z_{78}, t)$ from eqs. (6.7)

running, variation of the design and operating parameters can be studied to assess (a) the ALSS performance and (b) agreement of the model output with whatever experimental measurements are available.

The numerical methods used in ode_1 of Listing 6.2 and pde_1 of Listing 6.4 pertain to two main classes of PDEs:

- Eqs. (6.4) and (6.7) are first-order hyperbolic (convective) PDEs. Generally, this class of PDEs is relatively difficult to integrate numerically because they can propagate sharp fronts and discontinuities. In Listings 6.2 and 6.4, the convective terms were approximated by five-point biased upwind finite difference (FD) approximations in dss020, which generally perform quite well if the problem system has some smoothing. In the present case, smoothing is provided by the membrane transfer in MU1 and MU2 and by the adsorption in AU1 and AU2 so that the numerical solutions appear to be physically realistic (no numerical oscillations can be observed in Figs. 6.4). If unrealistic numerical artifacts are observed in the solutions, then a flux limiter would be required as discussed in Chapter 1.

- Eqs. (6.5a) and (6.6a) are second-order hyperbolic-parabolic (convective-diffusive) PDEs. Generally this class of PDEs has enough diffusion to smooth the solutions and thus avoid problems with steep fronts and discontinuities. A quantitative measure of the relative magnitudes of convection and diffusion is provided by the Péclet number, $Pe = (z_l)(v)/D$ where z_l is a characteristic length in the flow direction, v the flow velocity, and D the diffusivity in the direction of flow. For the present case, with the parameters taken from Listing 6.3, $z_{34} = 25, |v_3| = 2.5, D_3 = 0.01$, $Pe = (25)(2.5)/0.01 = 6250$. This is a large Pe indicating a strongly hyperbolic (convective) system, yet the FD approximations of dss020 appear to handle the moving fronts satisfactorily when used in combination with dss044 to include diffusion.

- Another consideration when numerically integrating second-order hyperbolic-parabolic PDEs is the requirement for a second exit or outflow BC (in addition to an inflow BC). In reported studies, this requirement is frequently satisfied by setting the slope (or first derivative in z) of the solution to zero. However, this is not a satisfactory BC since the solution cannot necessarily have a zero slope at the outflow point (consider a moving front solution). To circumvent this problem, dynamic outflow BCs [2] eqs. (6.5c) and (6.6c) are used that avoid the numerical distortions introduced by a zero-slope outflow BC. Experience with dynamic outflow BCs has generally been good and they are easily included in a MOL solution as demonstrated by eqs. (6.5) and (6.6).

- Since the MOL solutions provide the PDE dependent variables over a grid in z, additional insight into the solutions can be readily gained by plotting the spatial profiles of the solutions, e.g., $u_1(z,t)$ against z with t as a parameter. This type of analysis can also be extended by plotting the RHS terms of the PDEs against z with t as a parameter. In this way, the relative contributions of the RHS terms can be observed that generally model various phenomena of the physical system. Finally, the PDE LHS terms (derivatives in t) can be plotted against z with t as a parameter. All of this additional plotting is readily accomplished by using the computed numerical PDE solutions.

This discussion indicates that some experimentation with numerical methods, particularly for strongly hyperbolic (convective) PDEs, may be required. In addition to observing if numerical distortions such as unrealistic oscillations have occurred, an error analysis should be performed, for example, by changing the number of spatial grid points and observing the effects on the solution. This has not been reported here to conserve space, but a critical evaluation of the numerical solution should be part of any PDE modeling, and should not be overlooked because of the use of an available PDE solver.

APPENDIX - DERIVATION OF PDES FOR MEMBRANE AND ADSORPTION UNITS

The PDEs for the membrane and adsorption units in Fig. 6.3 are based on a toxin mass balance expressed first in words as eq. (A.6.1).

```
rate of toxin accumulation =                                      A.6.1a
        or depletion
  rate of convection into - rate of convection out  A.6.1b,c
      incremental volume       of incremental volume
+ rate of diffusion into   - rate of diffusion out   A.6.1d,e
      incremental volume       of incremental volume
+ rate of mass transfer across                       A.6.1f
  membrane or with adsorbent
```

The terms in eq. (A.6.1) are now considered for the membrane and adsorption units.

(A.6.1) PDEs FOR MEMBRANE UNITS

The PDEs are toxin balances for an incremental volume of the membrane unit of length Δz (z is the axial distance along a membrane or adsorption units as in Fig. 6.3). The membrane is considered as a bundle of hollow fibers or cylinders with a cross sectional area for flow and an axial or longitudinal area for mass transfer.

The terms in eq. (A.1) for MU1 in Fig. 6.3 are expressed mathematically.

- A.6.1a: $A_1 \Delta z \dfrac{\partial u_1}{\partial t}$; units: $(cm^2)(cm)(\mu mol/cm^3)(1/min)=\mu mol/min$
  where A_1 is the cross sectional area for membrane internal flow. The net units $\mu mol/min$ represent the accumulation or depletion of toxin per min within the incremental volume $A_1 \Delta z$.
- A.6.1b: $A_1 v_1 u_1|_z$; units: $(cm^2)(cm/min)(\mu mol/cm^3)=\mu mol/min$
  where v_1 is the linear flow velocity through the interior membrane. Again, the net units $\mu mol/min$ represent the convective flow into the incremental volume $A_1 \Delta z$ at z.
- A.6.1c: $-A_1 v_1 u_1|_{z+\Delta z}$; units: $(cm^2)(cm/min)(\mu mol/cm^3)=\mu mol/min$
  The net units $\mu mol/min$ represent the convective flow out of the incremental volume $A_1 \Delta z$ at $z + \Delta z$.

- **A.6.1d:** $-A_1 D_1 \frac{\partial u_1}{\partial z}|_z$; units: $(cm^2)(cm^2/min)(\mu mol/cm^3\text{-}cm)=\mu mol/min$

  The net units $\mu mol/min$ represent the diffusive flux into the incremental volume $A_1 \Delta z$ at z. D_1 is an effective diffusivity or axial dispersion coefficient. The minus in $-A_1 \cdots$ is required by Fick's first law to give a flux in the positive z direction when the gradient $\frac{\partial u_1}{\partial z}$ is negative (if this minus is overlooked, the numerical solution will most likely be unstable).

- **A.6.1e:** $A_1 D_1 \frac{\partial u_1}{\partial z}|_{z+\Delta z}$; units: $(cm^2)(cm^2/min)(\mu mol/cm^3\text{-}cm)=\mu mol/min$

  The net units $\mu mol/min$ represent the diffusive flux out of the incremental volume $A_1 \Delta z$ at $z + \Delta z$.

- **A.6.1f:** $A_{1m} \Delta z k_{12}(u_2 - u_1)$; units: $(cm^2/cm)(cm)(cm/min)(\mu mol/cm^3)= \mu mol/min$

  The net units $\mu mol/min$ represent the mass transfer rate between the internal and exterior membrane streams for the incremental length Δz. k_{12} is the mass transfer coefficient for the membrane and A_{1m} is the interior mass transfer area per unit length of membrane.

The consistency of units ($\mu mol/min$) in all of the terms in eq. (A.6.1) is an important requirement in deriving each PDE. Also, the consistency of the units for the parameters such as $A_1, v_1, D_1, A_{1m}, k_{12}$ is essential. As a result of the choice of units discussed above, z and t in $u_1(z, t)$ will be cm and min, respectively, and these units will be reflected in the numerical solution for u_1.

If the preceding terms are substituted in eq. (A.6.1), we obtain a mathematical form of this equation.

$$A_1 \Delta z \frac{\partial u_1}{\partial t} = A_1 v_1 u_1|_z - A_1 v_1 u_1|_{z+\Delta z}$$
$$- A_1 D_1 \frac{\partial u_1}{\partial z}|_z + A_1 D_1 \frac{\partial u_1}{\partial z}|_{z+\Delta z}$$
$$+ A_{1m} \Delta z k_{12}(u_2 - u_1) \qquad (A.6.2)$$

Division of eq. (A.6.2) by $A_1 \Delta z$ and minor rearrangement gives

$$\frac{\partial u_1}{\partial t} = -\frac{v_1 u_1|_{z+\Delta z} - v_1 u_1|_z}{\Delta z}$$
$$\frac{D_1 \frac{\partial u_1}{\partial z}|_{z+\Delta z} - D_1 \frac{\partial u_1}{\partial z}|_z}{\Delta z}$$
$$+ \frac{A_{1m}}{A_1} k_{12}(u_2 - u_1) \qquad (A.6.3)$$

With $\Delta z \to 0$, eq. (A.6.3) becomes

$$\frac{\partial u_1}{\partial t} = -\frac{\partial(v_1 u_1)}{\partial z} + \frac{\partial \left(D_1 \frac{\partial u_1}{\partial z} \right)}{\partial z} + \frac{A_{1m}}{A_1} k_{12}(u_2 - u_1) \qquad (A.6.4)$$

or for constant $v_1, D_1,$

$$\frac{\partial u_1}{\partial t} = -v_1 \frac{\partial u_1}{\partial z} + D_1 \frac{\partial^2 u_1}{\partial z^2} + \frac{A_{1m}}{A_1} k_{12}(u_2 - u_1) \qquad (A.6.5)$$

Eq. (A.6.5) is the PDE for the blood stream toxin concentration in Fig. 6.3, $u_1(z,t)$. Since it is second order in z and first order t, it requires two BCs and one IC. These auxiliary conditions are discussed earlier in this chapter.

$u_2(z,t)$, the albumin dialysate concentration, also requires a PDE. This follows directly from eq. (A.6.5).

$$\frac{\partial u_2}{\partial t} = -v_2 \frac{\partial u_2}{\partial z} + D_2 \frac{\partial^2 u_2}{\partial z^2} + \frac{A_{2m}}{A_2} k_{12}(u_1 - u_2) \qquad (A.6.6)$$

Eqs. (A.6.5) and (A.6.6) are the PDE model for the first membrane unit, MU1, in Fig. 6.3.

(A.6.2) PDEs FOR ADSORPTION UNITS

AU1 and AU2 in Fig. 6.3 are modeled in a similar fashion. The basic difference is that whereas MU1 and MU2 have two flowing streams, AU1 and AU2 have one flowing stream and one solid adsorbent phase. For AU1, the PDE toxin mass balances are based on the incremental volume $A_3 \Delta z$ where A_3 is the adsorber cross sectional area.

The PDE model for AU1 is

$$\frac{\partial u_3}{\partial t} = -v_3 \frac{\partial u_3}{\partial z} + D_3 \frac{\partial^2 u_3}{\partial z^2} - ((1-\epsilon)/\epsilon)k_{34}(u_4^e - u_4) \qquad (A.6.7)$$

The units of the variables and parameters in eq. (A.6.7) are explained below (μmol pertains to the toxin, $\mathrm{ml}_{fluid} + \mathrm{ml}_{adsorbent} = \mathrm{cm}^3_{adsorber}$).

u_3	μmol/ml$_{fluid}$
u_4	μmol/ml$_{adsorbent}$
u_4^e	μmol/ml$_{adsorbent}$
z	cm$_{adsorber}$
ϵ	ml$_{fluid}$/cm$^3_{adsorber}$
$1-\epsilon$	ml$_{adsorbent}$/cm$^3_{adsorber}$
v_3	cm$_{adsorber}$/min
D_3	cm$^2_{adsorber}$/min
k_{34}	1/min

The units of the terms in eq. (A.6.7) follow.

- $\dfrac{\partial u_3}{\partial t}$; units: ($\mu$mol/ml$_{fluid}$)(1/min)
  The units represent the rate of change of the toxin concentration from accumulation or depletion of the toxin.

- $-v_3\dfrac{\partial u_3}{\partial z}$; units: $(\text{cm}_{adsorber}/\text{min})(\mu\text{mol/ml}_{fluid})(1/\text{cm}_{adsorber})=(\mu\text{mol/ml}_{fluid})$ $(1/\text{min})$

  The net units represent the rate of change of the toxin concentration from convection of the toxin. v_3 is a fluid superficial velocity since it is based on $\text{cm}_{adsorber}$.

- $D_3\dfrac{\partial^2 u_3}{\partial z^2}$; units: $(\text{cm}^2_{adsorber}/\text{min})(\mu\text{mol/ml}_{fluid})(1/\text{cm}^2_{adsorber})=(\mu\text{mol/ml}_{fluid})$ $(1/\text{min})$

  The net units represent the rate of change of the toxin concentration from axial dispersion of the toxin.

- $-((1-\epsilon)/\epsilon)k_{34}(u_4^e - u_4)$; units: $(\text{ml}_{adsorbent}/\text{cm}^3_{adsorber})(\text{cm}^3_{adsorber}/\text{ml}_{fluid})$ $(1/\text{min})(\mu\text{mol/ml}_{adsorbent})=(\mu\text{mol/ml}_{fluid})(1/\text{min})$

  The net units represent the rate of change of the fluid toxin concentration u_3 from mass transfer of the toxin between the fluid and adsorbent. The rate of this mass transfer is given by $k_{34}(u_4^e - u_4)$. The equilibrium adsorbent concentration, u_4^e, is a function of u_3, generally in the form of an isotherm.

Eq. (A.6.7) is the PDE for the fluid stream toxin concentration in Fig. 6.3, $u_3(z,t)$. Since it is second order in z and first order t, it requires two BCs and one IC. These auxiliary conditions are discussed earlier in this chapter.

$u_4(z,t)$, the adsorbent concentration, also requires a PDE (to give $u_4(z,t)$ which is used in eq. (A.6.7)).

$$\frac{\partial u_4}{\partial t} = k_{34}(u_4^e - u_4) \tag{A.6.8}$$

The units of the two terms in eq. (A.6.8) are $(\mu\text{mol/ml}_{adsorbent})(1/\text{min})$. Eq. (A.6.8) does not have derivatives in z (the absorbent is not flowing or moving, and the axial diffusion along the adsorbent is considered negligible). Therefore, eq. (A.6.8) requires only one IC.

Eqs. (A.6.7) and (A.6.8) are the PDE model for the first adsorption unit, AU1, in Fig. 6.3. The PDEs for MU2 and AU2 in Fig. 6.3 follow in the same way as for MU1 and AU1.

A derivation of more general convection-diffusion-reaction PDEs in Cartesian, cylindrical and spherical coordinates is given in Appendix A.

References

[1] Annesni, M.C., V. Piemonte, and L. Turchetti (2014), Artificial liver support systems: a patient-device model, *Asia-Pacific J. Chem. Engr.*, **9**, 390–400; this reference has an excellent introduction to the function of the liver and an extensive list of references.

[2] Schiesser, W.E. (1996), PDE boundary conditions from minimum reduction of the PDE, *Appl. Numer. Math.*, **20**, 171–179.

7

CROSS DIFFUSION EPIDEMIOLOGY MODEL

In this chapter, an epidemiology model for the transmission of a disease is considered with the usual designation SIR (susceptible-infected-recovered individuals). The model has the following distinctive features [1]:

- It is a 2×2 system of nonlinear diffusion-reaction (parabolic) PDEs.
- The recovered individuals are considered to not have developed an immunity from the disease so that they can again become susceptible, that is, a SIS model.
- Cross diffusion in the PDEs is included to model how the infecteds can influence the movement of the susceptibles.

We start with a statement of the model PDEs, the initial conditions (ICs) and boundary conditions (BCs).

(7.1) 2-PDE model

The two PDEs of the model are taken from [1].

$$\frac{\partial u_1}{\partial t} = -\beta u_1^n u_2^m - bu_1 + \gamma u_2 + a(u_1 + u_2) + d_{11}\frac{\partial^2 u_1}{\partial x^2} + d_{12}\frac{\partial^2 u_2}{\partial x^2} \qquad (7.1a)$$

$$\frac{\partial u_2}{\partial t} = \beta u_1^n u_2^m - (\alpha + b + \gamma)u_2 + d_{22}\frac{\partial^2 u_2}{\partial x^2} \qquad (7.1b)$$

Method of Lines PDE Analysis in Biomedical Science and Engineering, First Edition. William E. Schiesser.
© 2016 John Wiley & Sons, Inc. Published 2016 by John Wiley & Sons, Inc.
Companion website: www.wiley.com/go/Schiesser/PDE_Analysis

Parameter	Numerical value
a	0.05
b	0.006
α	0.06
β	0.0056
γ	0.04
m	2
n	1
d_{11}	100
d_{12}	6
d_{22}	3
x_l	20

Table 7.1: Numerical values of the parameters in eqs. (7.1)

where

u_1	susceptibles concentration
u_2	infecteds concentration
x	spatial coordinate, $0 \leq x \leq x_l$
t	time, $t > 0$
m, n	constants for incidence rates of the disease
a	population birth rate
b	population death rate
γ	recovery rate
α	mortality cause by disease
β	disease transmission rate
d_{11}, d_{12}, d_{22}	diffusivities

The significance of the terms in eqs. (7.1) is considered subsequently. These equations are first order in t and second order in x so they each require one IC and two BCs.

$$u_1(x, t = 0) = g_1(t); \quad \frac{\partial u_1(x = 0, t)}{\partial x} = \frac{\partial u_1(x = x_l, t)}{\partial x} = 0 \qquad (7.2a,b,c)$$

$$u_2(x, t = 0) = g_2(t); \quad \frac{\partial u_2(x = 0, t)}{\partial x} = \frac{\partial u_2(x = x_l, t)}{\partial x} = 0 \qquad (7.2d,e,f)$$

The parameters in eqs. (7.1) are initially taken from [1], then subsequently redefined numerically to produce some particular features in the computed solutions.

To conclude this section on the model equations, the following physical interpretation of the terms in eqs. (7.1) provides some insight into the computed solutions discussed subsequently.

- Eq. (7.1a)
  1. $-\beta u_1^n u_2^m$: Rate of transmission of the disease from the susceptibles to the infecteds. This term reduces the susceptibles ($\beta > 0$). Also, both susceptibles and infecteds must be present for a nonzero rate.
  2. $-bu_1$: Natural death rate ($b > 0$) that reduces the susceptibles.
  3. $+\gamma u_2$: Recovery rate of infecteds that increases the susceptibles ($\gamma > 0$).
  4. $+a(u_1 + u_2)$: Rate of increase of susceptibles from susceptible and infected birth rates ($a > 0$).
  5. $+d_{11}\dfrac{\partial^2 u_1}{\partial x^2}$: Rate of increase of susceptibles by diffusion in the direction of decreasing susceptibles ($d_{11} > 0$) according to Fick's first law.
  6. $+d_{12}\dfrac{\partial^2 u_2}{\partial x^2}$: Rate of increase of susceptibles by diffusion in the direction of decreasing infecteds ($d_{12} > 0$). This is a cross diffusion term reflecting that susceptibles will move in the direction of decreasing infecteds.
  7. $\dfrac{\partial u_1}{\partial t}$: Rate of change of susceptibles with t from the combined effects of the preceding rates.
- Eq. (7.1b)
  1. $\beta u_1^n u_2^m$: Rate of transmission of the disease from the susceptibles to the infecteds. This term increases the infecteds ($\beta > 0$). Also, since this is the same term as in eq. (7.1a), but opposite in sign, the rate of increase of infecteds equals the rate of decrease of susceptibles.
  2. $-(\alpha + b + \gamma)u_2$: Rate of decrease of infecteds from the combined effects of (1) death by the disease ($\alpha > 0$), (2) natural deaths ($b > 0$) and (3) recovery from disease ($\gamma > 0$).
  3. $\dfrac{\partial u_2}{\partial t}$: Rate of change of susceptibles with t from the combined effects of the preceding rates.

(7.2) Model routines

Eqs. (7.1) and (7.2) constitute the epidemiology model that is programmed in the routines discussed next.

(7.2.1) Main program

The main program for the model is in Listing 7.1.

```
#
# Cross diffusion epidemiology model
#
# Delete previous work spaces
  rm(list=ls(all=TRUE))
#
# Access ODE integrator
  library("deSolve");
```

```
#
# Access functions for numerical solutions
  setwd("g:/chap7");
  source("pde_1.R");
  source("dss004.R");
  source("dss044.R");
#
# Level of output
#
#   ip = 1 - graphical (plotted) solutions
#             (u1(x,t), u2(x,t)) only
#
#   ip = 2 - numerical and graphical solutions
#
  ip=2;
#
# Parameters
  a=0.05;      b=0.006;
  alpha=0.06; beta=0.0056; gamma=0.04;
  m=2;         n=1;
  d11=100;     d12=6;        d22=3;
#
# Revised values
  d11=3; a=0.006; alpha=0.006; gamma=0.006;
#
# Select case
  ncase=1;
#
# Initial conditions
  nx=41;
  u0=rep(0,2*nx);
#
# ncase=1
  (if ncase==1){
    for(i in 1:nx){
      u0[i]=1;
      u0[i+nx]=0;
    }
    d12=0;
  }
#
# ncase=2,3
  if(ncase>1){
    for(i in 1:nx){
      u0[i]=1;
      if(i<=5){
```

```
        u0[i+nx]=1;
      }else{
        u0[i+nx]=0;
        }
      }
    if(ncase==2){d12=0;}
    if(ncase==3){d12=6;}
  }
  ncall=0;
#
# Write selected parameters
  cat(sprintf("\n\n ncase = %2d   d11 = %4.2f   d12 = %4.2f
             d22 = %4.2f", ncase,d11,d12,d22));
#
# Write heading
  if(ip==1){
    cat(sprintf("\n Graphical output only\n"));
  }
#
# Grid (in x)
  xl=0;xu=20;
  x=seq(from=xl,to=xu,by=(xu-xl)/(nx-1));
#
# Independent variable for ODE integration
  nout=5;t0=0;tf=20;
  tout=seq(from=t0,to=tf,by=(tf-t0)/(nout-1));
#
# ODE integration
  out=lsodes(y=u0,times=tout,func=pde_1,parms=NULL)
  nrow(out)
  ncol(out)
#
# Arrays for plotting numerical solution
  u1_plot=matrix(0,nrow=nx,ncol=nout);
  u2_plot=matrix(0,nrow=nx,ncol=nout);
  for(it in 1:nout){
    for(ix in 1:nx){
      u1_plot[ix,it]=out[it,ix+1];
      u2_plot[ix,it]=out[it,ix+1+nx];
    }
  }
#
# Display numerical solution
  if(ip==2){
    for(it in 1:nout){
```

```
      cat(sprintf(
      "\n    t    x    u1(x,t)      u2(x,t)\n"));
      for(ix in 1:nx){
        cat(sprintf("%5.1f%8.1f%12.3f%12.3f\n",
        tout[it],x[ix],u1_plot[ix,it],u2_plot[ix,it]));
      }
    }
  }
#
# Calls to ODE routine
  cat(sprintf("\n\n ncall = %5d\n\n",ncall));
#
# Plot u1
  par(mfrow=c(1,1));
  matplot(x=x,y=u1_plot,type="l",xlab="x",
          ylab="u1(x,t),  t=0,5,...,20",xlim=c(xl,xu),lty=1,
          main="u1(x,t);  t=0,5,...,20;",lwd=2);
#
# Plot u2
  par(mfrow=c(1,1));
  matplot(x=x,y=u2_plot,type="l",xlab="x",
          ylab="u2(x,t),  t=0,5,...,20",xlim=c(xl,xu),lty=1,
          main="u2(x,t);  t=0,5,...,20;",lwd=2);
```

Listing 7.1: Main program for eqs. (7.1) and (7.2)

We can note the following details about Listing 7.1.

- Previous work spaces are removed.

```
#
# Cross diffusion epidemiology model
#
# Delete previous work spaces
  rm(list=ls(all=TRUE))
```

- The ODE library deSolve is included (to access lsodes). Next, the files accessed by source include the ODE/MOL routine, pde_1, and two differentiation in space (DSS) routines, dss004, dss044.

```
#
# Access ODE integrator
  library("deSolve");
#
# Access functions for numerical solutions
  setwd("g:/chap7");
```

```
source("pde_1.R");
source("dss004.R");
source("dss044.R");
```

The setwd requires editing for the local computer. Note that / is used in place of the usual \.

- The level of numerical output is selected.

```
#
# Level of output
#
#   ip = 1 - graphical (plotted) solutions
#             (u1(x,t), u2(x,t)) only
#
#   ip = 2 - numerical and graphical solutions
#
   ip=2;
```

- The parameters in eqs. (7.1) are defined numerically. These values are from [1], then reset to reflect expected conditions, e.g., $d_{11} = 3$ in eq. (7.1a) to give a diffusivity that is in line with d_{12} (eq. (7.1a)) and d_{22} (eq. (7.1b)), $a = 0.006$ so that this birth rate is closer to the death rate ($b = 0.006$) to give a stable population.

```
#
# Parameters
  a=0.05;      b=0.006;
  alpha=0.06; beta=0.0056; gamma=0.04;
  m=2;         n=1;
  d11=100;    d12=6;        d22=3;
#
# Revised values
  d11=3; a=0.006; alpha=0.006; gamma=0.006;
```

- Three ICs are programmed as selected by ncase.

```
#
# Select case
  ncase=1;
#
# Initial conditions
  nx=41;
  u0=rep(0,2*nx);
#
# ncase=1
  (if ncase==1){
    for(i in 1:nx){
      u0[i]=1;
```

```
     u0[i+nx]=0;
    }
    d12=0;
  }
#
# ncase=2,3
  if(ncase>1){
    for(i in 1:nx){
      u0[i]=1;
      if(i<=5){
        u0[i+nx]=1;
      }else{
      u0[i+nx]=0;
      }
    }
    if(ncase==2){d12=0;}
    if(ncase==3){d12=6;}
  }
  ncall=0;
```

The ICs for ncase=1,2,3 are explained next and later in the discussion of the numerical and graphical output.

- ncase=1: $u_1(x, t = 0) = 1, u_2(x, t = 0) = 0$. The constant values for u_1, u_2 produce zero RHS terms in eqs. (7.1) so that the LHS derivatives in t are also zero and the solutions are constant at the ICs (as explained subsequently).
- ncase=2: $u_1(x, t = 0) = 1$ and $u_2(x \le 2, t = 0) = 1, u_2(x > 2, t = 0) = 0$ so that u_2 has a unit step change at $x = 2$. Cross diffusion is not included, d12=0. The model response to the discontinuous change in u_2 is discussed subsequently.
- ncase=3: $u_1(x, t = 0) = 1$ and $u_2(x \le 2, t = 0) = 1, u_2(x > 2, t = 0) = 0$ so that $u2$ again has a unit step change at $x = 2$. Cross diffusion is included, d12=6. The model response to the discontinuous change in u_2 with cross diffusion included is discussed subsequently.

The number of MOL/ODE ICs in u0 is $2(41) = 82$. The counter for the calls to the MOL/ODE routine is initialized.

- Since the three diffusivities d_{11}, d_{12}, d_{22} define the spatial effects of the model (variations with x), they are displayed in a summary at the beginning of the solution.

```
#
# Write selected parameters
  cat(sprintf("\n\n ncase = %2d   d11 = %4.2f   d12 = %4.2f
              d22 = %4.2f", ncase,d11,d12,d22));
#
# Write heading
  if(ip==1){
```

```
    cat(sprintf("\n Graphical output only\n"));
  }
```

The level of numerical output (with ip) is also confirmed.

- A grid in x is defined with the seq utility for $0 \leq x \leq 20$ with 41 points so $x = 0, 0.5, \ldots, 20$.

```
#
# Grid (in x)
  nx=41;xl=0;xu=20;
  x=seq(from=xl,to=xu,by=(xu-xl)/(nx-1));
```

- The output points in t are $t = 0, 5, 10, 15, 20$ (nout=5 values including $t = 0$).

```
#
# Independent variable for ODE integration
  nout=5;t0=0;tf=20;
  tout=seq(from=t0,to=tf,by=(tf-t0)/(nout-1));
```

- 82 ODEs are integrated by lsodes (the number of ODEs is provided to lsodes as the length of the IC vector u0).

```
#
# ODE integration
  out=lsodes(y=u0,times=tout,func=pde_1,parms=NULL)
  nrow(out)
  ncol(out)
```

The input arguments to lsodes are the IC vector, u0, the vector of output values of t, tout, the name of the MOL/ODE routine, pde_1 (discussed next), and parameters passed to pde_1 (unused). y,times,func,parms are reserved names. The output solution matrix is sized as out[nout,2(nz)+1]=out[5,83] which is confirmed by the nrow,ncol utilities (as reflected in the numerical output discussed subsequently).

- The numerical solutions of eqs. (7.1), u_1, u_2, are placed in arrays for subsequent plotting.

```
#
# Arrays for plotting numerical solution
  u1_plot=matrix(0,nrow=nx,ncol=nout);
  u2_plot=matrix(0,nrow=nx,ncol=nout);
  for(it in 1:nout){
    for(ix in 1:nx){
       u1_plot[ix,it]=out[it,ix+1];
       u2_plot[ix,it]=out[it,ix+1+nx];
    }
  }
```

The offset of 1 in [it,ix+1], [it,ix+1+nx] is required since out[it,1] has the output values of t.

- For ip=2, the numerical u_1, u_2 are displayed as a function of t (by a for with index it) and x (by a for with index ix).

```
#
# Display numerical solution
  if(ip==2){
    for(it in 1:nout){
      cat(sprintf(
      "\n    t    x     u1(x,t)     u2(x,t)\n"));
      for(ix in 1:nx){
        cat(sprintf("%5.1f%8.1f%12.3f%12.3f\n",
        tout[it],x[ix],u1_plot[ix,it],u2_plot[ix,it]));
      }
    }
  }
```

- The number of calls to pde_1 is displayed at the end of the solution as an indication of the computational effort to compute the solution (by lsodes).

```
#
# Calls to ODE routine
  cat(sprintf("\n\n ncall = %5d\n\n",ncall));
```

- The numerical solutions, u_1, u_2, are plotted against x with t as a parameter by the matplot utility. Scaling in the vertical direction is done automatically.

```
#
# Plot u1
  par(mfrow=c(1,1));
  matplot(x=x,y=u1_plot,type="l",xlab="x",
          ylab="u1(x,t), t=0,5,...,20",xlim=c(xl,xu),lty=1,
          main="u1(x,t); t=0,5,...,20;",lwd=2);
#
# Plot u2
  par(mfrow=c(1,1));
  matplot(x=x,y=u2_plot,type="l",xlab="x",
          ylab="u2(x,t), t=0,5,...,20",xlim=c(xl,xu),lty=1,
          main="u2(x,t); t=0,5,...,20;",lwd=2);
```

lsodes calls the MOL/ODE routine pde_1 listed next.

(7.2.2) ODE routine

The MOL/ODE routine follows.

```
  pde_1=function(t,u,parms){
#
# Function pde_1 computes the t derivative vector of the
# u1, u2 vectors
#
# One vector to two vectors
  u1=rep(0,nx);u2=rep(0,nx);
  for(i in 1:nx){
    u1[i]=u[i];
    u2[i]=u[i+nx];
  }
#
# u1x, u2x
  u1x=dss004(xl,xu,nx,u1);
  u2x=dss004(xl,xu,nx,u2);
#
# Boundary conditions
  u1x[1]=0;u1x[nx]=0;
  u2x[1]=0;u2x[nx]=0;
  nl=2;nu=2;
#
# u1xx, u2xx
  u1xx=dss044(xl,xu,nx,u1,u1x,nl,nu);
  u2xx=dss044(xl,xu,nx,u2,u2x,nl,nu);
#
# Nonlinear term
  u1u2=rep(0,nx);
  for(i in 1:nx){
    u1u2[i]=u1[i]^n*u2[i]^m;
  }
#
# PDEs
  u1t=rep(0,nx);u2t=rep(0,nx);
  for(i in 1:nx){
    u1t[i]=-beta*u1u2[i]-b*u1[i]+gamma*u2[i]+
              a*(u1[i]+u2[i])+d11*u1xx[i]+
              d12*u2xx[i];
    u2t[i]= beta*u1u2[i]-(alpha+b+gamma)*u2[i]+
              d22*u2xx[i];
  }
#
# Two vectors to one vector
```

```
  ut=rep(0,2*nx);
  for(i in 1:nx){
    ut[i]      =u1t[i];
    ut[i+nx]   =u2t[i];
  }
#
# Increment calls to pde_1
  ncall <<- ncall+1;
#
# Return derivative vector
  return(list(c(ut)));
}
```

Listing 7.2: MOL/ODE routine for eqs. (7.1)

We can note the following details about Listing 7.2.

- The function is defined. u is an input 82-vector of the ODE dependent variables at $t = $ t. parms for passing parameters to pde_1 is unused.

  ```
  pde_1=function(t,u,parms){
  #
  # Function pde_1 computes the t derivative vector of the
  # u1, u2 vectors
  ```

- u is placed in two vectors, u1, u2, to facilitate the programming of eqs. (7.1). These vectors are first declared with the rep utility. nx=41 is set numerically in the main program of Listing 7.1 and passed to pde_1 without any special designation (a feature of R).

  ```
  #
  # One vector to two vectors
    u1=rep(0,nx);u2=rep(0,nx);
    for(i in 1:nx){
      u1[i]=u[i];
      u2[i]=u[i+nx];
    }
  ```

- The first derivatives $\partial u_1/\partial x$, $\partial u_2/\partial x$ in eqs. (7.1) are computed by the library differentiator dss004.

  ```
  #
  # u1x, u2x
    u1x=dss004(xl,xu,nx,u1);
    u2x=dss004(xl,xu,nx,u2);
  ```

- BCs (7.2b,c,e,f) are defined numerically. Note the use of subscripts 1,nx corresponding to $x = 0, 20$.

```
#
# Boundary conditions
  u1x[1]=0;u1x[nx]=0;
  u2x[1]=0;u2x[nx]=0;
  nl=2;nu=2;
```

nl=nu=2 specifies Neumann BCs since the first derivatives are defined (rather than the dependent variables u_1, u_2 at the boundaries which would be specified with nl=nu=1).

- The second derivatives $\partial^2 u_1/\partial x^2$, $\partial^2 u_2/\partial x^2$ in eqs. (7.1) are computed by the library differentiator dss044.

```
#
# u1xx, u2xx
  u1xx=dss044(xl,xu,nx,u1,u1x,nl,nu);
  u2xx=dss044(xl,xu,nx,u2,u2x,nl,nu);
```

- The nonlinear term $u_1^n u_2^m$ is computed and placed in vector u1u2.

```
#
# Nonlinear term
  u1u2=rep(0,nx);
  for(i in 1:nx){
    u1u2[i]=u1[i]^n*u2[i]^m;
  }
```

This programming illustrates the ease of including nonlinearities in a numerical solution.

- Eqs. (7.1) are programmed. The derivatives $\partial u_1/\partial t$, $\partial u_2/\partial t$ are placed in vectors u1t,u2t.

```
#
# PDEs
  u1t=rep(0,nx);u2t=rep(0,nx);
  for(i in 1:nx){
    u1t[i]=-beta*u1u2[i]-b*u1[i]+gamma*u2[i]+
            a*(u1[i]+u2[i])+d11*u1xx[i]+
            d12*u2xx[i];
    u2t[i]= beta*u1u2[i]-(alpha+b+gamma)*u2[i]+
            d22*u2xx[i];
  }
```

The close correspondence of this programming with eqs. (7.1) illustrates an important advantage of the MOL solution of PDEs.

- Vectors u1t,u2t are placed in a single vector ut for return to lsodes (called in Listing 7.1).

```
#
# Two vectors to one vector
  ut=rep(0,2*nx);
  for(i in 1:nx){
    ut[i]      =u1t[i];
    ut[i+nx]   =u2t[i];
  }
```

- The number of calls to pde_1 is incremented and the value returned to the main program of Listing 7.1 by a <<-.

```
#
# Increment calls to pde_1
  ncall <<- ncall+1;
```

- The derivative vector ut is returned to lsodes by a combination of c, the vector operator, list (lsodes requires a list), and return.

```
#
# Return derivative vector
  return(list(c(ut)));
}
```

The final } completes pde_1.

This completes the programming of eqs. (7.1) and (7.2). The numerical and graphical output from the main program of Listing 7.1 is considered next.

(7.3) Model output

The output for ncase=1,2,3 is considered in the following discussion.

(7.3.1) ncase $= 1$, time-invariant solution

We can note the following details about this output.

- The dimensions of the solution matrix out are confirmed as 5×83 (based on the previous discussion).
- The IC for $t = 0$ is confirmed (as programmed in Listing 7.1 for ncase=1).
- The solution remains at the IC. This invariance in t results from all the RHS terms in eqs. (7.1) equal to zero, that is,

```
ncase =  1   d11 = 3.00   d12 = 0.00   d22 = 3.00
[1] 5
[1] 83
       t        x      u1(x,t)      u2(x,t)
      0.0      0.0      1.000       0.000
      0.0      0.5      1.000       0.000
      0.0      1.0      1.000       0.000
      0.0      1.5      1.000       0.000
      0.0      2.0      1.000       0.000
                .         .
                .         .
                .         .
Output from x = 2.5 to 17.5 removed
                .         .
                .         .
                .         .
      0.0     18.0      1.000       0.000
      0.0     18.5      1.000       0.000
      0.0     19.0      1.000       0.000
      0.0     19.5      1.000       0.000
      0.0     20.0      1.000       0.000
                .         .
                .         .
                .         .
Output for t = 5, 10, 15 removed
                .         .
                .         .
                .         .
       t        x      u1(x,t)      u2(x,t)
     20.0      0.0      1.000       0.000
     20.0      0.5      1.000       0.000
     20.0      1.0      1.000       0.000
     20.0      1.5      1.000       0.000
     20.0      2.0      1.000       0.000
                .         .
                .         .
                .         .
Output from x = 2.5 to 17.5 removed
                .         .
                .         .
                .         .
```

Table 7.2: Abbreviated output for **ncase=1**

20.0	18.0	1.000	0.000
20.0	18.5	1.000	0.000
20.0	19.0	1.000	0.000
20.0	19.5	1.000	0.000
20.0	20.0	1.000	0.000

ncall =   102

Table 7.2: (*Continued*)

1. $-\beta u_1^n u_2^m = 0$ with $u_2 = 0$.

2. $-bu_1 + \gamma u_2 + a(u_1 + u_2) = 0$ with $a = b, u_2 = 0$.

3. $d_{11}\dfrac{\partial^2 u_1}{\partial x^2} = 0$ with $u_1 = 1$ (a constant).

4. $d_{12}\dfrac{\partial^2 u_2}{\partial x^2} = 0$ with $u_2 = 0$ (a constant).

5. $\beta u_1^n u_2^m = 0$ with $u_2 = 0$.

6. $-(\alpha + b + \gamma)u_2 = 0$ with $u_2 = 0$.

7. $d_{22}\dfrac{\partial^2 u_2}{\partial x^2} = 0$ with $u_2 = 0$ (a constant).

- The computational effort is small, ncall = 102, as expected since the solution does not change with t (or x).

The graphical output (plotting) confirms the output in Table 7.2 and therefore is not presented here.

This case is worth running since if the solution changed with t, this would indicate a programming error.

(7.3.2) ncase = 2, transient solution, no cross diffusion

Abbreviated numerical output is in Table 7.3.

We can note the following details about this output.

- The dimensions of the solution matrix out are again confirmed as 5×83 (based on the previous discussion).
- The IC for $t = 0$ is confirmed (as programmed in Listing 7.1 for ncase=2). Note the step in u_2 at $x = 2$.
- The solution is a response to the step in u_2, as displayed in Figs. 7.1a, 7.1b, with no cross diffusion since d12=0.
- The computational effort is modest, ncall = 277, even with the discontinuous change in u_2, which is to be expected since eqs. (7.1) represent a form of diffusion that tends to smooth the solution with increasing t, as demonstrated in Fig. 7.1b.

```
ncase =  2   d11 = 3.00   d12 = 0.00   d22 = 3.00
[1]  5
[1]  83
        t        x       u1(x,t)      u2(x,t)
       0.0      0.0       1.000        1.000
       0.0      0.5       1.000        1.000
       0.0      1.0       1.000        1.000
       0.0      1.5       1.000        1.000
       0.0      2.0       1.000        1.000
       0.0      2.5       1.000        0.000
       0.0      3.0       1.000        0.000
       0.0      3.5       1.000        0.000
       0.0      4.0       1.000        0.000
                  .          .
                  .          .
                  .          .
                  .          .
Output from x = 4.5 to 17.5 removed
                  .          .
                  .          .
                  .          .
       0.0     18.0       1.000        0.000
       0.0     18.5       1.000        0.000
       0.0     19.0       1.000        0.000
       0.0     19.5       1.000        0.000
       0.0     20.0       1.000        0.000
                  .          .
                  .          .
                  .          .
Output for t = 5, 10, 15 removed
                  .          .
                  .          .
                  .          .
        t        x       u1(x,t)      u2(x,t)
      20.0      0.0       1.030        0.117
      20.0      0.5       1.030        0.117
      20.0      1.0       1.030        0.116
      20.0      1.5       1.030        0.116
      20.0      2.0       1.030        0.115
                  .          .
                  .          .
                  .          .
```

Table 7.3: Abbreviated output for ncase=2

```
Output from x = 2.5 to 17.5 removed

         .         .
         .         .
         .         .
20.0    18.0      1.012        0.046
20.0    18.5      1.012        0.046
20.0    19.0      1.012        0.045
20.0    19.5      1.012        0.045
20.0    20.0      1.012        0.045

ncall =    277
```

Table 7.3: (*Continued*)

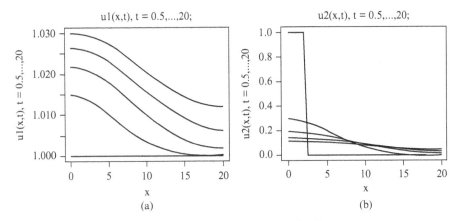

Figure 7.1a   (a) $u_1(x,t)$ for ncase=2, (b) $u_2(x,t)$ for ncase=2

(7.3.3) ncase $= 3$, transient solution with cross diffusion

Abbreviated numerical output is in Table 7.4.

  We can note the following details about this output.

- The dimensions of the solution matrix out are again confirmed as 5×83 (based on the previous discussion).
- Cross diffusion has been added with d12=6.
- The IC for $t = 0$ is confirmed (as programmed in Listing 7.1 for ncase=3). Note the step in u_2 at $x = 2$.
- The solution is a response to the step in u_2, as displayed in Figs. 7.2a, 7.2b, with cross diffusion since d12=6.
- The computational effort is modest, ncall = 299.

```
ncase =  3   d11 = 3.00   d12 = 6.00   d22 = 3.00
[1] 5
[1] 83
```

t	x	u1(x,t)	u2(x,t)
0.0	0.0	1.000	1.000
0.0	0.5	1.000	1.000
0.0	1.0	1.000	1.000
0.0	1.5	1.000	1.000
0.0	2.0	1.000	1.000
0.0	2.5	1.000	0.000
0.0	3.0	1.000	0.000
0.0	3.5	1.000	0.000
0.0	4.0	1.000	0.000
.	.		
.	.		
.	.		

Output from x = 4.5 to 17.5 removed

t	x	u1(x,t)	u2(x,t)
.	.		
.	.		
.	.		
0.0	18.0	1.000	0.000
0.0	18.5	1.000	0.000
0.0	19.0	1.000	0.000
0.0	19.5	1.000	0.000
0.0	20.0	1.000	0.000
.	.		
.	.		
.	.		

Output for t = 5, 10, 15 removed

t	x	u1(x,t)	u2(x,t)
.	.		
.	.		
.	.		
20.0	0.0	0.898	0.116
20.0	0.5	0.899	0.116
20.0	1.0	0.900	0.116
20.0	1.5	0.902	0.115
20.0	2.0	0.905	0.114
20.0	2.5	0.909	0.113
20.0	3.0	0.914	0.112
20.0	3.5	0.919	0.111
20.0	4.0	0.925	0.109

Table 7.4: Abbreviated output for ncase=3

```
                .                .
                .                .
                .                .
   Output from x = 2.5 to 17.5 removed
                .                .
                .                .
                .                .
   20.0      18.0           1.128           0.046
   20.0      18.5           1.130           0.046
   20.0      19.0           1.132           0.045
   20.0      19.5           1.133           0.045
   20.0      20.0           1.133           0.045

   ncall =    299
```

Table 7.4: (*Continued*)

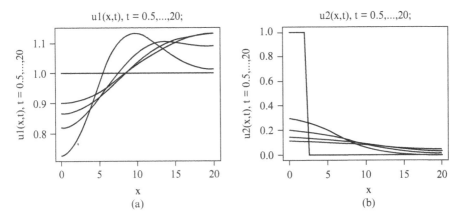

Figure 7.2   (a) $u_1(x,t)$ for ncase=3, (b) $u_2(x,t)$ for ncase=3

Note that the solution for u_1 has changed substantially (comparing Figs. 7.1a and 7.1b) as a result of adding cross diffusion (changing d12=0 for ncase=2 to d12=6 for ncase=3).

Again, as in Fig. 7.1b, diffusion smoothes the IC unit step.

(7.4) Summary and conclusions

The preceding analysis of eqs. (7.1) and (7.2) indicates that cross diffusion is easily accommodated numerically (e.g., in Listing 7.2), and can have a major effect on the solution of the model PDEs.

Also, the discussion for ncase=1 illustrates how the individual terms in the PDEs can be analyzed in detail, first the RHS terms, then the LHS terms that are the derivatives in t and that determine how the solution evolves in t. This procedure of analyzing and comparing the individual terms can be generalized to any case by computing and plotting the terms (as a function of t and x). All that is required to do this is to have the numerical solutions (e.g., u_1, u_2) to compute the terms. In this way, we can obtain a detailed picture of how the principal features of the solution are produced, and this in turn can lead to a directed evaluation and possible revision of the model. In other words, the contribution of the terms that model the physical and chemical phenomena of the problem system can be evaluated and modified during the development of the model.

Reference

[1] Aly, S., H.B. Khenous, and F. Hussien (2015), Turing instability in a diffusive SIS epidemi-
ological model, *Int. J. Biomath.*, **8**, no. 1.

8

ONCOLYTIC VIROTHERAPY

This chapter pertains to a mathematical model for cancer therapy based on the use of viruses, that is, oncolytic virotherapy (immunotherapy, immuno-oncology). The basic concepts of the therapy are described in the following excerpt taken from Malinzi et al, [1].

> Oncolytic virotherapy uses replication-competent viruses to kill cancer cells. Specific viruses are turned into therapeutic agents to treat cancer. The idea of using viruses as a treatment for cancer began in the 1950s, when tissue culture and rodent cancer models were originally developed. Today, oncolytic treatment involves the use of virus genomes which are engineered to enhance their anti-tumor specificity. This began with a study in which thymidine kinase-negative HSV with attenuated neurovirulence was shown to be active in a murine glioblastoma model. Since then, the pace of clinical activities has accelerated considerably, with several trials using oncolytic viruses belonging to different virus families. To date, clinical trials have pointed out talimogene laherparepvec as a possible treatment for melanoma. Clinical trials are ongoing using different viruses as cancer therapies. There is no recorded toxicity as a result of clinical use of oncolytic virotherapy to treat cancer (references included in 1, p 102, not included here).

In other words, the following model demonstrates *immunotherapy* in which the immune system is used against cancer cells. The special case of using viruses to activate the immune system is virotherapy.

The intent of this chapter is to:

- Present a 4-PDE model for oncolytic virotherapy including the required initial conditions (ICs) and boundary conditions (BCs).

Method of Lines PDE Analysis in Biomedical Science and Engineering, First Edition. William E. Schiesser.
© 2016 John Wiley & Sons, Inc. Published 2016 by John Wiley & Sons, Inc.
Companion website: www.wiley.com/go/Schiesser/PDE_Analysis

- Discuss the format of the model as a nonlinear *diffusion-reaction* (or *parabolic*) PDE system in 1D.
- Illustrate an incremental approach to the coding of the model in which RHS terms are added one at a time to the PDEs, and the routines are executed to demonstrate the effect of the added terms. The method of lines (MOL) solution includes the use of library routines for integration of the PDE derivatives in time and space.
- Present the computed model solution in numerical and graphical (plotted) format.
- Discuss the features of the numerical solution and the performance of the algorithms used to compute the solution. In particular, two sources of numerical error are discussed.
  - The effect of a spatial grid that has too few points to achieve accurate spatial resolution of the numerical solution.
  - The effect of a discontinuity in the RHS of a PDE.

The 4-PDE oncolytic virotherapy model reported in [1] is presented next, followed by a method of lines (MOL) solution programmed in R.

(8.1) 1D 4-PDE model

The PDE model is

$$
\frac{\partial u_1}{\partial t} = D_1 \left(\frac{\partial^2 u_1}{\partial r^2} + \frac{2}{r} \frac{\partial u_1}{\partial r} \right) + \lambda \frac{\partial}{\partial r} \left(u_1 \frac{\partial u_4}{\partial r} \right)
$$

$$
+ \sigma h(r) - \psi u_1 + \frac{\gamma_1 u_1 u_3}{\eta_1 + u_3} - \nu u_1 u_3 \tag{8.1a}
$$

$$
\frac{\partial u_2}{\partial t} = D_2 \left(\frac{\partial^2 u_2}{\partial r^2} + \frac{2}{r} \frac{\partial u_2}{\partial r} \right) + \alpha_1 u_2 (1 - \alpha_2 u_2) - \frac{\theta_1 u_2 u_3}{\eta_2 + u_3} \tag{8.1b}
$$

$$
\frac{\partial u_3}{\partial t} = D_3 \left(\frac{\partial^2 u_3}{\partial r^2} + \frac{2}{r} \frac{\partial u_3}{\partial r} \right) + \beta_1 u_3 (1 - \beta_2 u_3) + \frac{\theta_2 u_2 u_3}{\eta_2 + u_3} - \mu u_1 u_3 \tag{8.1c}
$$

$$
\frac{\partial u_4}{\partial t} = D_4 \left(\frac{\partial^2 u_4}{\partial r^2} + \frac{2}{r} \frac{\partial u_4}{\partial r} \right) + \frac{\gamma_2 u_1 u_3}{\eta_1 + u_3} - \delta u_4 \tag{8.1d}
$$

where

Variable	Meaning
u_1	cytotoxic-T-cell (CTL) density
u_2	uninfected tumor cell density
u_3	infected tumor cell density
u_4	chemokine concentration
r	spatial (boundary value) variable (distance within the tumor)
t	time (initial value) variable
$h(t)$	unit step (Heaviside) function

Eqs. (8.1) are stated in *spherical coordinates*, (r, ϕ, θ), (note the differential group $\partial^2/\partial r^2 + (2/r)\partial/\partial r$) with no angular variation or terms in ϕ, θ from symmetry.
Eqs. (8.1) are first order in t are therefore require one IC for each PDE.

$$u_1(r, t=0) = \begin{cases} 0, & 0 \le r < r_b \\ u_{10}\left[1 - \exp\left(-100(r - r_b)^2\right)\right], & r_b \le r \le 1 \end{cases} \tag{8.2a}$$

$$u_2(r, t=0) = \begin{cases} u_{20}\left[1 - \exp\left(-100(r - r_b)^2\right)\right], & 0 \le r < r_b \\ 0, & r_b \le r \le 1 \end{cases} \tag{8.2b}$$

$$u_3(r, t=0) = \begin{cases} u_{30}\left[1 - \exp\left(-100(r - r_b)^2\right)\right], & 0 \le r < r_b \\ 0, & r_b \le r \le 1 \end{cases} \tag{8.2c}$$

$$u_4(r, t=0) = \begin{cases} 0, & 0 \le r < r_b \\ u_{40}\left[1 - \exp\left(-100(r - r_b)^2\right)\right], & r_b \le r \le 1 \end{cases} \tag{8.2d}$$

ICs (8.2) are a Gaussian distribution[1] (exponentials of the form e^{-cr^2}) defined over the domain $0 \le r \le 1$.
Eqs. (8.1) are second order in r and they therefore require two BCs for each PDE.

$$\frac{\partial u_1(r=0, t)}{\partial r} = \frac{\partial u_1(r=1, t)}{\partial r} = 0 \tag{8.3a,b}$$

$$\frac{\partial u_2(r=0, t)}{\partial r} = \frac{\partial u_2(r=1, t)}{\partial r} = 0 \tag{8.3c,d}$$

$$\frac{\partial u_3(r=0, t)}{\partial r} = \frac{\partial u_3(r=1, t)}{\partial r} = 0 \tag{8.3e,f}$$

$$\frac{\partial u_4(r=0, t)}{\partial r} = \frac{\partial u_4(r=1, t)}{\partial r} = 0 \tag{8.3g,h}$$

Eqs. (8.3) are homogeneous (zero) *Neumann* BCs since they specify the first derivatives in r. Physically, they can be interpreted as no variation of u_1, u_2, u_3, u_4 with r because of (a) symmetry at $r = 0$ and (b) no diffusion from the spherical domain $0 \le r \le 1$ at $r = 1$.
Numerical values are assigned to the parameters (constants) of eqs. (8.1) and (8.2) in the R routines presented and discussed next.

(8.2) MOL routines

The following routines implement the MOL solution of eqs. (8.1), (8.2) and (8.3). The main program is discussed first.

---

[1] An important advantage of the Gaussian function is that it is smooth, including all of its derivatives in r. This property facilitates the numerical integration of eqs. (8.1), which is also enhanced by the parabolic form of the PDEs (they model diffusion that smoothes the numerical solution) in contrast with the hyperbolic PDEs discussed in Chapter 1.

(8.2.1) Main program

The intent of this main program is to demonstrate how the PDEs are built or formulated by adding terms for various physical/chemical/biological phenomena. As new terms are added, their effect can be assessed by comparison of the numerical solution with the preceding solutions (without the terms)[2].

```
# Delete previous workspaces
  rm(list=ls(all=TRUE))
#
#   1D, 4-PDE oncolytic virotherapy model
#
# Access ODE integrator
  library("deSolve");
#
# Access files
  setwd("g:/chap8");
  source("pde_1.R") ;source("h.R");
  source("dss004.R");
#
# Select case (each case is explained below)
#
  ncase=1;
#
# Model parameters
    nr=41;   r0=1;
    u10=1; u20=1; u30=1; u40=0;
     D1=0;   D2=0;   D3=0;   D4=0;
    alpha1=0;alpha2=0;
    beta1=0;beta2=0;
    delta=0;
    eta1=1;eta2=1;
    gam1=0;gam2=0;
    lam=0;
    mu=0;
    nu=0;
    psi=0
    sigma=0;
    theta1=0;theta2=0;
#
```

---

[2]This incremental approach is intended to demonstrate how a mathematical model can be constructed and tested. This step-by-step approach is recommended for relatively complex problem systems, as opposed to including all of the contemplated or anticipated terms at the beginning of the PDE analysis to reflect the various physical/chemical/biological phenomena that are considered important. That is, adding one term at a time demonstrates the effect of that term, and if the model unexpectedly stops executing, the last term added is probably the cause of this failure to produce a solution.

```
# No change after IC
  if(ncase==1){
  }
#
# Diffusion only
  if(ncase==2){
    D1=0.01;D2=0.01;
    D3=0.01;D4=0.01;
  }
#
# Diffusion, tumor cell growth
  if(ncase==3){
     D1=0.01; D2=0.01;
     D3=0.01; D4=0.01;
    alpha1=1;alpha2=1;
     beta1=1; beta2=1;
  }
#
# Diffusion, tumor cell growth, chemotaxis
  if(ncase==4){
     D1=0.01; D2=0.01;
     D3=0.01; D4=0.01;
    alpha1=1;alpha2=1;
     beta1=1; beta2=1;
       lam=1;   u40=1;
  }
#
# Diffusion, tumor cell growth, chemotaxis,
# CTL-chemokine depletion
  if(ncase==5){
     D1=0.01; D2=0.01;
     D3=0.01; D4=0.01;
    alpha1=1;alpha2=1;
     beta1=1; beta2=1;
       lam=1;   u40=1;
     delta=1;   psi=1;
  }
#
# Diffusion, tumor cell growth, chemotaxis,
# CTL-chemokine depletion, CTL proliferation
  if(ncase==6){
     D1=0.01; D2=0.01;
     D3=0.01; D4=0.01;
    alpha1=1;alpha2=1;
     beta1=1; beta2=1;
       lam=1;   u40=1;
```

```
       delta=1;    psi=1;
        gam1=1;   gam2=1;
   }
#
# Diffusion, tumor cell growth, chemotaxis,
# CTL-chemokine depletion, CTL proliferation,
# virus replication
   if(ncase==7){
       D1=0.01; D2=0.01;
       D3=0.01; D4=0.01;
     alpha1=1;alpha2=1;
      beta1=1; beta2=1;
        lam=1;    u40=1;
      delta=1;    psi=1;
       gam1=1;   gam2=1;
      theta1=1;theta2=1;
   }
#
# Diffusion, tumor cell growth, chemotaxis,
# CTL-chemokine depletion, CTL proliferation,
# virus replication, immune cell supply
   if(ncase==8){
       D1=0.01; D2=0.01;
       D3=0.01; D4=0.01;
     alpha1=1;alpha2=1;
      beta1=1; beta2=1;
        lam=1;    u40=1;
      delta=1;    psi=1;
       gam1=1;   gam2=1;
      theta1=1;theta2=1;
      sigma=1;
   }
#
# Diffusion, tumor cell growth, chemotaxis,
# CTL-chemokine depletion, CTL proliferation,
# virus replication, immune cell supply,
# tumor cell death
   if(ncase==9){
       D1=0.01; D2=0.01;
       D3=0.01; D4=0.01;
     alpha1=1;alpha2=1;
      beta1=1; beta2=1;
        lam=1;    u40=1;
      delta=1;    psi=1;
       gam1=1;   gam2=1;
      theta1=1;theta2=1;
```

```
    sigma=1;
       nu=1;    mu=1;
  }
#
# Level of output
#
#   Detailed numerical and graphical output - ip = 1
#
#   Graphical output only - ip = 2
#
  ip=1;
#
# Spatial grid
  r=seq(from=0,to=r0,by=(r0-0)/(nr-1));
#
# Initial condition
  u0=rep(0,4*nr);rb=0.5
#
# u1(r,t=0)
  for(i in 1:nr){
    if(r[i]<rb){
      u0[i]=0;
    }else{
      u0[i]=u10*(1-exp(-100*(r[i]-rb)^{2}));
    }
  }
#
# u2(r,t=0)
  for(i in 1:nr){
    if(r[i]<rb){
      u0[nr+i]=u20*(1-exp(-100*(r[i]-rb)^{2}));
    }else{
      u0[nr+i]=0;
    }
  }
#
# u3(r,t=0)
  for(i in 1:nr){
    if(r[i]<rb){
      u0[2*nr+i]=u30*(1-exp(-100*(r[i]-rb)^{2}));
    }else{
      u0[2*nr+i]=0;
    }
  }
#
# u4(r,t=0)
```

```
  for(i in 1:nr){
    if(r[i]<rb){
      u0[3*nr+i]=0;
    }else{
      u0[3*nr+i]=u40*(1-exp(-100*(r[i]-rb)^{2})));
    }
  }
NROW(u0)
NCOL(u0)
#
# Grid in t
  t0=0;tf=2;nout=5;
  tout=seq(from=t0,to=tf,by=(tf-t0)/(nout-1));
  ncall=0;
#
# ODE integration
  out=ode(func=pde_1,times=tout,y=u0);
#
# Store solution
  u1=matrix(0,nrow=nout,ncol=nr);
  u2=matrix(0,nrow=nout,ncol=nr);
  u3=matrix(0,nrow=nout,ncol=nr);
  u4=matrix(0,nrow=nout,ncol=nr);
   t=rep(0,nout);
  for(it in 1:nout){
    for(i in 1:nr){
      u1[it,i]=out[it,i+1];
      u2[it,i]=out[it,nr+i+1];
      u3[it,i]=out[it,2*nr+i+1];
      u4[it,i]=out[it,3*nr+i+1];
    }
    t[it]=out[it,1];
  }
#
# Display  ncase, ncall
  cat(sprintf("\n\n ncase = %2d    ncall = %2d",
              ncase,ncall));
#
# Display numerical solution
  if(ip==1){
    for(it in 1:nout){
    cat(sprintf("\n\n t = %3.1f\n",t[it]));
    cat(sprintf(
      "\n    r    u1(r,t)    u2(r,t)    u3(r,t)    u4(r,t)\n"));
    for(i  in 1:nr  ){
      cat(sprintf(
```

```
          "%4.3f%10.4f%10.4f%10.4f%10.4f\n",
          r[i],u1[it,i],u2[it,i],u3[it,i],u4[it,i]));
    }
    }
  }
#
# Plot u1, u2, u3, u4
#
# u1(r,t)
  par(mfrow=c(1,1));
  matplot(r,t(u1),type="l",xlab="r",ylab="u1(r,t)",
    main="u1(r,t) vs r",col="black",lwd=2,lty=1);
#
# u2(r,t)
  par(mfrow=c(1,1));
  matplot(r,t(u2),type="l",xlab="r",ylab="u2(r,t)",
    main="u2(r,t) vs r",col="black",lwd=2,lty=1);
#
# u3(r,t)
  par(mfrow=c(1,1));
  matplot(r,t(u3),type="l",xlab="r",ylab="u3(r,t)",
    main="u3(r,t) vs r",col="black",lwd=2,lty=1);
#
# u4(r,t)
  par(mfrow=c(1,1));
  matplot(r,t(u4),type="l",xlab="r",ylab="u4(r,t)",
    main="u4(r,t) vs r",col="black",lwd=2,lty=1);
```

Listing 8.1: Main program for the solution of eqs. (8.1), (8.2) and (8.3)

We can note the following details about this main program.

- Previous workspaces are cleared.

```
#
# Delete previous workspaces
  rm(list=ls(all=TRUE))
#
#    1D, 4-PDE oncolytic virotherapy model
```

- The R library of ODE integrators, deSolve, and the routines for the MOL solution of eqs. (8.1) to (8.3) are accessed. Note that in the setwd (set working directory), / is used rather than the usual \.

```
#
# Access ODE integrator
  library("deSolve");
```

```
#
# Access files
  setwd("g:/chap8");
  source("pde_1.R") ;source("h.R");
  source("dss004.R");
```

- Nine cases are programmed corresponding to the incremental addition of RHS terms to eqs. (8.1).

```
#
# Select case (each case is explained below)
#
  ncase=1;
```

First, a set of base case parameters is defined numerically.

```
#
# Model parameters
    nr=41;   r0=1;
    u10=1; u20=1; u30=1; u40=0;
     D1=0;   D2=0;   D3=0;   D4=0;
    alpha1=0;alpha2=0;
    beta1=0;beta2=0;
    delta=0;
    eta1=1;eta2=1;
    gam1=0;gam2=0;
    lam=0;
    mu=0;
    nu=0;
    psi=0
    sigma=0;
    theta1=0;theta2=0;
```

Then, a particular case is selected with redefinition of the parameters for that case.

```
#
# No change after IC
  if(ncase==1){
  }
#
# Diffusion only
  if(ncase==2){
    D1=0.01;D2=0.01;
    D3=0.01;D4=0.01;
  }
         .    .
         .    .
         .    .
```

Here, for ncase=2, diffusion is added to the base case of ncase=1.

- A level of numerical output is selected.

```
#
# Level of output
#
#    Detailed numerical and graphical output - ip = 1
#
#    Graphical output only - ip = 2
#
   ip=1;
```

- A grid in r is defined for nr=41 points (nr is defined in the base case parameters) over the interval $0 \le r \le r_0$ with $r_0 = 1$ using the R utility seq.

```
#
# Spatial grid
   r=seq(from=0,to=r0,by=(r0-0)/(nr-1));
```

- ICs (8.2) are defined numerically.

```
#
# Initial condition
   u0=rep(0,4*nr);rb=0.5
#
# u1(r,t=0)
   for(i in 1:nr){
     if(r[i]<rb){
       u0[i]=0;
     }else{
       u0[i]=u10*(1-exp(-100*(r[i]-rb)^{2}));
     }
   }

                 .              .
                 .              .
                 .              .

#
# u4(r,t=0)
   for(i in 1:nr){
     if(r[i]<rb){
       u0[3*nr+i]=0;
     }else{
       u0[3*nr+i]=u40*(1-exp(-100*(r[i]-rb)^{2}));
     }
   }
NROW(u0)
NCOL(u0)
```

The IC vector u0 has 4*41 = 164 elements. r_b in eqs. (8.2) is set to rb=0.5. The four Gaussian distributions in eqs. (8.2) are defined for $u_1(r, t = 0)$, $u_2(r, t = 0)$, $u_3(r, t = 0)$, $u_4(r, t = 0)$.

The utilities NROW, NCOL are used to confirm the dimensions of u0. Note these names are capitalized for application to a vector (rather than a matrix) as explained in help(NROW), help(NCOL) entered at the R prompt.

- A grid in t is defined for the interval $0 \le t \le 2$ with nout=5 output points, $t = 0, 0.5, \ldots, 2$, using the seq utility.

```
#
# Grid in t
  t0=0;tf=2;nout=5;
  tout=seq(from=t0,to=tf,by=(tf-t0)/(nout-1));
  ncall=0;
```

- The 164 ODEs are programmed in pde_1 (discussed next) and integrated by ode (from deSolve). The ODE routine pde_1, the vector of output values tout and the IC vector u0 are the input to ode as expected (to define the MOL/ODE system). func, times, y are reserved names for ode.

```
#
# ODE integration
  out=ode(func=pde_1,times=tout,y=u0);
```

- The numerical solution in out is placed in four arrays, u1, u2, u3, u4. The solution values for the interval in t are included in the matrices by a for with index it, and for the interval in r by a for with index i. The offset of 1 in the second subscript is required since out[it,1] has the output values of t (placed in t[it]).

```
#
# Store solution
  u1=matrix(0,nrow=nout,ncol=nr);
  u2=matrix(0,nrow=nout,ncol=nr);
  u3=matrix(0,nrow=nout,ncol=nr);
  u4=matrix(0,nrow=nout,ncol=nr);
   t=rep(0,nout);
  for(it in 1:nout){
    for(i in 1:nr){
      u1[it,i]=out[it,i+1];
      u2[it,i]=out[it,nr+i+1];
      u3[it,i]=out[it,2*nr+i+1];
      u4[it,i]=out[it,3*nr+i+1];
    }
    t[it]=out[it,1];
  }
```

- Selected parameters are displayed at the beginning of the numerical solution.

```
#
# Display  ncase, ncall
  cat(sprintf("\n\n ncase = %2d    ncall = %2d",
              ncase,ncall));
```

- For ip=1 (detailed numerical output), the solutions of eqs. (8.1), $u_1(r,t)$, $u_2(r,t)$, $u_3(r,t)$, $u_4(r,t)$, are displayed as a function of t with the outer for in it and as a function of r with the inner for in i.

```
#
# Display numerical solution
  if(ip==1){
    for(it in 1:nout){
    cat(sprintf("\n\n t = %3.1f\n",t[it]));
    cat(sprintf(
      "\n    r    u1(r,t)   u2(r,t)   u3(r,t)   u4(r,t)\n"));
    for(i  in 1:nr  ){
      cat(sprintf(
        "%4.3f%10.4f%10.4f%10.4f%10.4f\n",
        r[i],u1[it,i],u2[it,i],u3[it,i],u4[it,i]));
    }
    }
  }
```

- Four plots of $u_1(r,t)$, $u_2(r,t)$, $u_3(r,t)$, $u_4(r,t)$ against r with $t = 0, 0.5, \ldots 2$ as a parameter are produced with the matplot utility.

```
#
# Plot u1, u2, u3, u4
#
# u1(r,t)
  par(mfrow=c(1,1));
  matplot(r,t(u1),type="l",xlab="r",ylab="u1(r,t)",
    main="u1(r,t) vs r",col="black",lwd=2,lty=1);
#
# u2(r,t)
  par(mfrow=c(1,1));
  matplot(r,t(u2),type="l",xlab="r",ylab="u2(r,t)",
    main="u2(r,t) vs r",col="black",lwd=2,lty=1);
#
# u3(r,t)
  par(mfrow=c(1,1));
  matplot(r,t(u3),type="l",xlab="r",ylab="u3(r,t)",
    main="u3(r,t) vs r",col="black",lwd=2,lty=1);
```

```
  #
  # u4(r,t)
    par(mfrow=c(1,1));
    matplot(r,t(u4),type="l",xlab="r",ylab="u4(r,t)",
```

The MOL/ODE routine called by the main program of Listing 8.1 that has the programming of eqs. (8.1) and (8.3) is considered next.

(8.2.2) MOL/ODE routine

pde_1 is in Listing 8.2.

```
  pde_1=function(t,u,parms) {
#
# Function pde_1 computes the t derivative vector
# of the u vector
#
# One vector to four PDE vectors
  u1=rep(0,nr);u2=rep(0,nr);
  u3=rep(0,nr);u4=rep(0,nr);
  for (i in 1:nr){
    u1[i]=u[i];
    u2[i]=u[nr+i];
    u3[i]=u[2*nr+i];
    u4[i]=u[3*nr+i];
  }
#
# u1r,u2r,u3r,u4r
  u1r=dss004(0,r0,nr,u1);
  u2r=dss004(0,r0,nr,u2);
  u3r=dss004(0,r0,nr,u3);
  u4r=dss004(0,r0,nr,u4);
#
# Boundary conditions
  u1r[1]=0;u1r[nr]=0;
  u2r[1]=0;u2r[nr]=0;
  u3r[1]=0;u3r[nr]=0;
  u4r[1]=0;u4r[nr]=0;
#
# u1rr,u2rr,u3rr,u4rr
  u1rr=dss004(0,r0,nr,u1r);
  u2rr=dss004(0,r0,nr,u2r);
  u3rr=dss004(0,r0,nr,u3r);
  u4rr=dss004(0,r0,nr,u4r);
#
# Chemotaxis (ct)
```

```
  uct=rep(0,nr);
  for(i in 1:nr){
    uct[i]=u1[i]*u4r[i];
  }
  uctr=dss004(0,r0,nr,uct);
#
# Temporal derivatives
    u1t=rep(0,nr);u2t=rep(0,nr);
    u3t=rep(0,nr);u4t=rep(0,nr);
#
#   u1t, u2t, u3t, u4t
    for(i in 1:nr){
      if(i==1){
        u1t[i]=D1*3*u1rr[i]+
               lam*uctr[i]+sigma*h(r[i])-psi*u1[i]+
               (gam1*u1[i]*u3[i])/(eta1+u3[i])-
               nu*u1[i]*u3[i];
        u2t[i]=D2*3*u2rr[i]+
               alpha1*u2[i]*(1-alpha2*u2[i])-
               theta1*u2[i]*u3[i]/(eta2+u3[i]);
        u3t[i]=D3*3*u3rr[i]+
               beta1*u3[i]*(1-beta2*u3[i])+
               theta2*u2[i]*u3[i]/(eta2+u3[i])-
               mu*u1[i]*u3[i];
        u4t[i]=D4*3*u4rr[i]+
               gam2*u1[i]*u3[i]/(eta1+u3[i])-
               delta*u4[i];
      }else{
        u1t[i]=D1*(u1rr[i]+2/r[i]*u1r[i])+
               lam*uctr[i]+sigma*h(r[i])-psi*u1[i]+
               (gam1*u1[i]*u3[i])/(eta1+u3[i])-
               nu*u1[i]*u3[i];
        u2t[i]=D2*(u2rr[i]+2/r[i]*u2r[i])+
               alpha1*u2[i]*(1-alpha2*u2[i])-
               theta1*u2[i]*u3[i]/(eta2+u3[i]);
        u3t[i]=D3*(u3rr[i]+2/r[i]*u3r[i])+
               beta1*u3[i]*(1-beta2*u3[i])+
               theta2*u2[i]*u3[i]/(eta2+u3[i])-
               mu*u1[i]*u3[i];
        u4t[i]=D4*(u4rr[i]+2/r[i]*u4r[i])+
               gam2*u1[i]*u3[i]/(eta1+u3[i])-
               delta*u4[i];
      }
    }
#
# Four PDE derivative vectors to one vector
```

```
  ut=rep(0,4*nr);
  for (i in 1:nr){
    ut[i]     =u1t[i];
    ut[nr+i]  =u2t[i];
    ut[2*nr+i]=u3t[i];
    ut[3*nr+i]=u4t[i];
  }
#
# Increment calls to pde_1
  ncall<<-ncall+1;
#
# Return derivative vector
  return(list(c(ut)));
}
```

Listing 8.2: pde_1 for eqs. (8.1) and (8.3)

We can note the following details about Listing 8.2.

- The function is defined.

```
    pde_1=function(t,u,parms) {
  #
  # Function pde_1 computes the t derivative vector
  # of the u vector
```

  u is the 164-vector of MOL/ODE dependent variables at the current value of t. parms, for passing parameters to pde_1, is unused, but requires designation as the third input argument.

- u is placed in four 41-vectors, u1,u2,u3,u4, to facilitate the programming of eqs. (8.1) in terms of problem oriented variables. These four vectors are first declared (preallocated) with the rep utility.

```
  #
  # One vector to four PDE vectors
    u1=rep(0,nr);u2=rep(0,nr);
    u3=rep(0,nr);u4=rep(0,nr);
    for (i in 1:nr){
      u1[i]=u[i];
      u2[i]=u[nr+i];
      u3[i]=u[2*nr+i];
      u4[i]=u[3*nr+i];
    }
```

- The first derivatives in r in eqs. (8.1), $\partial u_1/\partial r, \ldots, \partial u_4/\partial r$, are computed with the spatial differentiation routine dss004. The arrays for these derivatives, u1r, u2r, u3r, u4r, do not have to be declared since this is done in dss004.

```
#
# u1r,u2r,u3r,u4r
  u1r=dss004(0,r0,nr,u1);
  u2r=dss004(0,r0,nr,u2);
  u3r=dss004(0,r0,nr,u3);
  u4r=dss004(0,r0,nr,u4);
```

- BCs (8.3) are implemented by redefining the r derivatives at the boundaries $r = 0, 1$. Note the use of subscripts 1, nr corresponding to $r = 0, 1$.

```
#
# Boundary conditions
  u1r[1]=0;u1r[nr]=0;
  u2r[1]=0;u2r[nr]=0;
  u3r[1]=0;u3r[nr]=0;
  u4r[1]=0;u4r[nr]=0;
```

- The second derivatives in r in eqs. (8.1), $\partial^2 u_1/\partial r^2, \ldots, \partial^2 u_4/\partial r^2$, are computed with the spatial differentiation routine dss004 by differentiating the first derivatives, termed *stagewise differentiation*.

```
#
# u1rr,u2rr,u3rr,u4rr
  u1rr=dss004(0,r0,nr,u1r);
  u2rr=dss004(0,r0,nr,u2r);
  u3rr=dss004(0,r0,nr,u3r);
  u4rr=dss004(0,r0,nr,u4r);
```

- The chemotaxis term in eq. (8.1a), $\dfrac{\partial}{\partial r}\left(u_1 \dfrac{\partial u_4}{\partial r}\right)$, is computed in two steps: (1) the product $u_1 \dfrac{\partial u_4}{\partial r}$ is formed first in a for, and (2) this product is then differentiated with dss004. This procedure demonstrates how nonlinearities can be easily accommodated numerically.

```
#
# Chemotaxis (ct)
  uct=rep(0,nr);
  for(i in 1:nr){
    uct[i]=u1[i]*u4r[i];
  }
  uctr=dss004(0,r0,nr,uct);
```

- All of the RHS terms of eqs. (8.1) are now available so they can be combined to give the LHS terms (derivatives in t). These derivatives are first declared as four 41-vectors.

```
#
# Temporal derivatives
    u1t=rep(0,nr);u2t=rep(0,nr);
    u3t=rep(0,nr);u4t=rep(0,nr);
```

- The RHS of each PDE is then computed. For example, for eq. (8.1a), the programming for $r = 0$ is (with i=1 and r[1]=0 from the definition of r in the main program of Listing 8.1)

```
u1t[i]=D1*3*u1rr[i]+
       lam*uctr[i]+sigma*h(r[i])-psi*u1[i]+
       (gam1*u1[i]*u3[i])/(eta1+u3[i])-
       nu*u1[i]*u3[i];
```

and for $r > 0$ (i=2,...,nr and r[i]$\neq$0)

```
u1t[i]=D1*(u1rr[i]+2/r[i]*u1r[i])+
       lam*uctr[i]+sigma*h(r[i])-psi*u1[i]+
       (gam1*u1[i]*u3[i])/(eta1+u3[i])-
       nu*u1[i]*u3[i];
```

The special case at $r = 0$ is required because the radial group $\dfrac{2}{r}\dfrac{\partial u_1}{\partial r}$ in eq. (8.1a) is indeterminate ($0/0$, considering BC (8.3a)). This indeterminate term is regularized by the application of *l'Hospital's rule*.

$$\lim_{r\to 0}\frac{2}{r}\frac{\partial u_1}{\partial r} = \lim_{r\to 0} 2\frac{\partial^2 u_1}{\partial r^2}$$

so that

$$\lim_{r\to 0}\frac{\partial^2 u_1}{\partial r^2} + \frac{2}{r}\frac{\partial u_1}{\partial r} = \lim_{r\to 0} 3\frac{\partial^2 u_1}{\partial r^2}$$

(the factor 3 appears in the preceding coding of eq. (8.1a) for $r = 0$). Also, the similarity of the PDE (eq. (8.1a)) and the coding is clear, which is a principal advantage of the MOL. h(r[i]) is a call to a routine for the Heaviside function (unit step) considered subsequently. All of the parameters, such as D1,lam,sigma, etc., are defined numerically in the main program of Listing 8.1 and passed to pde_1 of Listing 8.2 without any special designation, which is a feature of R.

The programming of eqs. (8.1b) to (8.1d) follows in a similar way (in Listing 8.2), including the special case for $r = 0$ (i=1).

- The $4 \times 41 = 164$ derivatives in u1t,u2t,u3t,u4t are placed in a single derivative vector ut for return to the ODE integrator ode.

```
#
# Four PDE derivative vectors to one vector
  ut=rep(0,4*nr);
  for (i in 1:nr){
    ut[i]      =u1t[i];
    ut[nr+i]   =u2t[i];
    ut[2*nr+i]=u3t[i];
    ut[3*nr+i]=u4t[i];
  }
```

- The number of calls to pde_1 is incremented and returned to the main program of Listing 8.1 by the operator <<-.

```
#
# Increment calls to pde_1
  ncall<<-ncall+1;
```

- The derivative vector ut is returned to the ODE integrator by the combination of (1) c, the vector operator in R, (2) list (ode requires a list) and (3) return.

```
#
# Return derivative vector
  return(list(c(ut)));
}
```

The final } concludes pde_1.

The overall objective in executing pde_1 of Listing 8.2 is to compute the 164 ODE derivatives in ut corresponding to the 164 dependent variables coming into pde_1 as input argument u. The relative positions of the dependent variables and their derivatives must be maintained, that is, the derivative ut[i] must correspond to dependent variable u[i], i=1,2,...,164. The expectation is that ode will integrate ut forward in t to give u at the next output value of t in tout. If this forward integration takes place through all of the output values of t in tout, a complete solution to eqs. (8.1) to (8.3) will result that can be observed numerically and graphically.

Before we consider a solution, the remaining function for discussion is h.

(8.2.3) Subordinate routine

h for the Heaviside (unit step) function in eq. (8.1a) is listed next.

```
  h=function(r){
#
# Function h defines the unit step (Heaviside) function
# for the supply of immune cells
#
  rb=0.5;
  if(r<rb){
    h=0;
  }else{
    h=1;
  }
#
# Return function
  return(c(h));
}
```

Listing 8.3: h for the Heaviside function in eq. (8.1a)

We can note the following details of Listing 8.3.

- The function is defined.

```
h=function(r){
#
# Function h defines the unit step (Heaviside) function
# for the supply of immune cells
```

- For $r < r_b$ (r_b is used in ICs (8.2)), the function is zero.

```
#
rb=0.5;
if(r<rb){
   h=0;
```

The value of r_b is 0.5 rather than 0.2 in [1] to center the step change in $0 \le r \le 1$, and is also defined in the main program of Listing 8.1.

- Otherwise, for $r \ge r_b$, the function is one.

```
}else{
   h=1;
}
```

- The value of the function is returned as a vector through a combination of c (the vector operator applied to the scalar h), and return.

```
#
# Return function
   return(c(h));
}
```

The final } concludes h.

The programming of eqs. (8.1) to (8.3) is now complete and we can consider the numerical solution.

(8.3) Model output

The main program of Listing 8.1 produces extensive numerical and graphical (plotted) output. Here we consider only selected output to conserve space, but the reader can easily execute the routines to generate the full output. The discussion follows through the successive values ncase=1,2,...,9 in Listing 8.1.

- For ncase=1, the base case parameters in Listing 8.1 set all of the RHS terms in eqs. (8.1) to zero. Thus, all of the LHS derivatives in t are also zero and therefore the solution should not change from the ICs of eqs. (8.2). This case may seem obvious and trivial, but it is worth executing since if the solution moves away from the IC, there is an error in the coding (probably in pde_1 of Listing 8.2). Abbreviated numerical output for this case is given in Table 8.1.

```
[1] 164
[1] 1
 ncase =  1    ncall =   8
 t = 0.0
     r    u1(r,t)     u2(r,t)     u3(r,t)     u4(r,t)
0.000    0.0000      1.0000      1.0000      0.0000
0.025    0.0000      1.0000      1.0000      0.0000
0.050    0.0000      1.0000      1.0000      0.0000
0.075    0.0000      1.0000      1.0000      0.0000
0.100    0.0000      1.0000      1.0000      0.0000
0.125    0.0000      1.0000      1.0000      0.0000
0.150    0.0000      1.0000      1.0000      0.0000
0.175    0.0000      1.0000      1.0000      0.0000
0.200    0.0000      0.9999      0.9999      0.0000
0.225    0.0000      0.9995      0.9995      0.0000
0.250    0.0000      0.9981      0.9981      0.0000
0.275    0.0000      0.9937      0.9937      0.0000
0.300    0.0000      0.9817      0.9817      0.0000
0.325    0.0000      0.9532      0.9532      0.0000
0.350    0.0000      0.8946      0.8946      0.0000
0.375    0.0000      0.7904      0.7904      0.0000
0.400    0.0000      0.6321      0.6321      0.0000
0.425    0.0000      0.4302      0.4302      0.0000
0.450    0.0000      0.2212      0.2212      0.0000
0.475    0.0000      0.0606      0.0606      0.0000
0.500    0.0000      0.0000      0.0000      0.0000
0.525    0.0606      0.0000      0.0000      0.0000
0.550    0.2212      0.0000      0.0000      0.0000
0.575    0.4302      0.0000      0.0000      0.0000
0.600    0.6321      0.0000      0.0000      0.0000
0.625    0.7904      0.0000      0.0000      0.0000
0.650    0.8946      0.0000      0.0000      0.0000
0.675    0.9532      0.0000      0.0000      0.0000
0.700    0.9817      0.0000      0.0000      0.0000
0.725    0.9937      0.0000      0.0000      0.0000
0.750    0.9981      0.0000      0.0000      0.0000
0.775    0.9995      0.0000      0.0000      0.0000
0.800    0.9999      0.0000      0.0000      0.0000
0.825    1.0000      0.0000      0.0000      0.0000
0.850    1.0000      0.0000      0.0000      0.0000
0.875    1.0000      0.0000      0.0000      0.0000
0.900    1.0000      0.0000      0.0000      0.0000
0.925    1.0000      0.0000      0.0000      0.0000
0.950    1.0000      0.0000      0.0000      0.0000
0.975    1.0000      0.0000      0.0000      0.0000
```

Table 8.1: Abbreviated numerical output, ncase=1

1.000   1.0000   0.0000   0.0000   0.0000

.   .

.   .

.   .

Output for t = 0.5, 1, 1.5 removed

.   .

.   .

.   .

t = 2.0

r	u1(r,t)	u2(r,t)	u3(r,t)	u4(r,t)
0.000	0.0000	1.0000	1.0000	0.0000
0.025	0.0000	1.0000	1.0000	0.0000
0.050	0.0000	1.0000	1.0000	0.0000
0.075	0.0000	1.0000	1.0000	0.0000
0.100	0.0000	1.0000	1.0000	0.0000
0.125	0.0000	1.0000	1.0000	0.0000
0.150	0.0000	1.0000	1.0000	0.0000
0.175	0.0000	1.0000	1.0000	0.0000
0.200	0.0000	0.9999	0.9999	0.0000
0.225	0.0000	0.9995	0.9995	0.0000
0.250	0.0000	0.9981	0.9981	0.0000
0.275	0.0000	0.9937	0.9937	0.0000
0.300	0.0000	0.9817	0.9817	0.0000
0.325	0.0000	0.9532	0.9532	0.0000
0.350	0.0000	0.8946	0.8946	0.0000
0.375	0.0000	0.7904	0.7904	0.0000
0.400	0.0000	0.6321	0.6321	0.0000
0.425	0.0000	0.4302	0.4302	0.0000
0.450	0.0000	0.2212	0.2212	0.0000
0.475	0.0000	0.0606	0.0606	0.0000
0.500	0.0000	0.0000	0.0000	0.0000
0.525	0.0606	0.0000	0.0000	0.0000
0.550	0.2212	0.0000	0.0000	0.0000
0.575	0.4302	0.0000	0.0000	0.0000
0.600	0.6321	0.0000	0.0000	0.0000
0.625	0.7904	0.0000	0.0000	0.0000
0.650	0.8946	0.0000	0.0000	0.0000
0.675	0.9532	0.0000	0.0000	0.0000
0.700	0.9817	0.0000	0.0000	0.0000
0.725	0.9937	0.0000	0.0000	0.0000
0.750	0.9981	0.0000	0.0000	0.0000
0.775	0.9995	0.0000	0.0000	0.0000
0.800	0.9999	0.0000	0.0000	0.0000
0.825	1.0000	0.0000	0.0000	0.0000
0.850	1.0000	0.0000	0.0000	0.0000

Table 8.1: (*Continued*)

0.875	1.0000	0.0000	0.0000	0.0000
0.900	1.0000	0.0000	0.0000	0.0000
0.925	1.0000	0.0000	0.0000	0.0000
0.950	1.0000	0.0000	0.0000	0.0000
0.975	1.0000	0.0000	0.0000	0.0000
1.000	1.0000	0.0000	0.0000	0.0000

Table 8.1: (*Continued*)

We can note the following details of this output.

- The IC array u0 (Listing 8.1) is a 164×1 column vector as expected. The length of the vector informs ode of the number of ODEs to be integrated (since u0 is an input to ode in Listing 8.1).

- ncase=1 is confirmed, and the number of calls to pde_1 (Listing 8.2) indicates that the ODE integration in t required essentially no computations.

```
ncase =   1    ncall =   8
```

- The ICs of eqs. (8.2) (t = 0.0) are confirmed in the sense that (1) the Gaussian distributions are identical for u_1, u_2, u_3 with the exception of displacements in t, and (2) the IC for u_4 is zero (since u40=0). This may seem like an obvious conclusion, but checking the ICs is always a good idea, since if they are incorrect, the subsequent solution will also be incorrect.

- The solution remains at the IC, as expected since both sides of eqs. (8.1) are zero. This is confirmed at $t = 2$ (as well as at $t = 0.5, 1, 1.5$ which are not included to conserve space).

The t-invariant solution is confirmed by the graphical output (from Listing 8.1). For each execution, four plots are produced for u_1, u_2, u_3, u_4 as a function of r with t as a parameter. Thus, for ncase=1,2,...,9, $4 \times 9 = 36$ plots would result, too many for the available space. Thus, we will consider the graphical output for a subset of plots for u_1, u_2, u_3, u_4 in each case, with the choice determined by the change in the selected dependent variable. For ncase=1, the plot of u_3 (concentration of infected cells) is in Fig. 8.1, which indicates that the solution remains at the IC (five curves are superimposed and are indistinguishable).

- For ncase=2, diffusion of the four components with concentrations u_1, u_2, u_3, u_4 is added. This is accomplished by replacing the zero diffusivities of the base case with nonzero values in eqs. (8.1), that is, in the terms

$$D_1 \left(\frac{\partial^2 u_1}{\partial r^2} + \frac{2}{r} \frac{\partial u_1}{\partial r} \right), \dots, D_4 \left(\frac{\partial^2 u_4}{\partial r^2} + \frac{2}{r} \frac{\partial u_1}{\partial r} \right).$$

```
#
# Diffusion only
  if(ncase==2){
    D1=0.01;D2=0.01;
    D3=0.01;D4=0.01;
  }
```

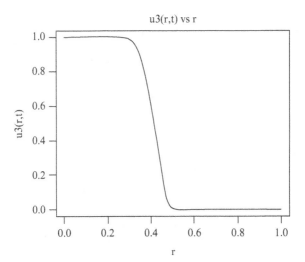

Figure 8.1 $u_3(r, t)$ vs r with $t = 0, 0.5, \ldots, 2$ as a parameter for `ncase=1`

We would now expect the Gaussian distributions to be smoothed in r and t. This is confirmed in the following numerical and graphical output.

We can note the following details of this output.

- The IC array u0 (Listing 8.1) is the same as for `ncase=1` in Table 8.1, as expected.
- `ncase=2` is confirmed, and the number of calls to `pde_1` (Listing 8.2) is `ncall = 765` which is modest.

```
ncase =   2    ncall = 765
```

- The ICs of eqs. (8.2) are smoothed by diffusion, except for u_4, which remains at the IC of zero since the RHS diffusion term in eq. (8.1d) does not move u_4 from this zero IC.
- The solutions for u_1, u_2, u_3, u_4 at least qualitatively follow BCs (8.3) with little change in r at the boundaries $r = 0, 1$. In other words, the ICs of eqs. (8.2) and the BCs of (8.3) are consistent (or compatible).

This final point of a zero slope at the boundaries is confirmed by the graphical output in Fig. 8.2 (again for $u_3(r, t)$).

The solution in Fig. 8.2 demonstrates a general smoothing of the IC Gaussian function through $t = 0, 0.5, \ldots, 2$ and for large t ($t \gg 2$) will approach a uniform concentration in r as expected (from diffusion).

- For `ncase=3`, tumor cell growth is added to diffusion. This is done by changing the zero values of $\alpha_1, \alpha_2, \beta_1, \beta_2$ in eqs. (8.1b) and (8.1c) to nonzero values in the terms $\alpha_1 u_2(1 - \alpha_2 u_2)$, $\beta_1 u_3(1 - \beta_2 u_3)$.

```
#
# Diffusion, tumor cell growth
  if(ncase==3){
```

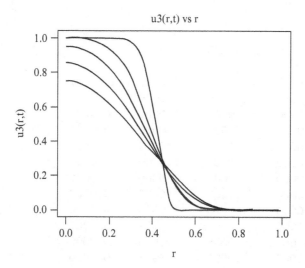

Figure 8.2 $u_3(r, t)$ vs r with $t = 0, 0.5, \ldots, 2$ as a parameter for `ncase=2`

```
  D1=0.01; D2=0.01;
  D3=0.01; D4=0.01;
  alpha1=1;alpha2=1;
   beta1=1; beta2=1;
}
```

We can observe the effect of $\alpha_1, \alpha_2, \beta_1, \beta_2$ by comparing the solutions for `ncase=2` in Table 8.2 and Fig. 8.2 with the solution in the following Table 8.3 and Fig. 8.3. We can note the following details of this output.

– `ncase=3` is confirmed, and the number of calls to `pde_1` (Listing 8.2) is `ncall` = 772 which is modest.

```
  ncase =   3    ncall = 772
```

– The ICs of eqs. (8.2) are the same as for `ncase=1,2` and therefore are not included in Table 8.3 to conserve space (as well as the output for $t = 0.5, 1, 1.5$).

– The solutions for u_2, u_3 are greater than for `ncase=2` since tumor cell growth (uninfected, u_2, and infected, u_3) has been added. For example, for $r = 0.5$, $t = 2$,

```
ncase = 2 (Table 8.2)
0.500    0.4752    0.1915    0.1915    0.0000
ncase = 3 (Table 8.3)
0.500    0.4752    0.4899    0.4899    0.0000
```

Thus, the effect of adding nonzero values of $\alpha_1, \alpha_2, \beta_1, \beta_2$ is as expected (increased tumor cell concentrations) and is confirmed by the step-by-step procedure in going from ncase=2 to ncase=3.

```
[1] 164
[1] 1
 ncase =  2    ncall = 765
 t = 0.0
```

r	u1(r,t)	u2(r,t)	u3(r,t)	u4(r,t)
0.000	0.0000	1.0000	1.0000	0.0000
0.025	0.0000	1.0000	1.0000	0.0000
0.050	0.0000	1.0000	1.0000	0.0000
0.075	0.0000	1.0000	1.0000	0.0000
0.100	0.0000	1.0000	1.0000	0.0000
0.125	0.0000	1.0000	1.0000	0.0000
0.150	0.0000	1.0000	1.0000	0.0000
0.175	0.0000	1.0000	1.0000	0.0000
0.200	0.0000	0.9999	0.9999	0.0000
0.225	0.0000	0.9995	0.9995	0.0000
0.250	0.0000	0.9981	0.9981	0.0000
0.275	0.0000	0.9937	0.9937	0.0000
0.300	0.0000	0.9817	0.9817	0.0000
0.325	0.0000	0.9532	0.9532	0.0000
0.350	0.0000	0.8946	0.8946	0.0000
0.375	0.0000	0.7904	0.7904	0.0000
0.400	0.0000	0.6321	0.6321	0.0000
0.425	0.0000	0.4302	0.4302	0.0000
0.450	0.0000	0.2212	0.2212	0.0000
0.475	0.0000	0.0606	0.0606	0.0000
0.500	0.0000	0.0000	0.0000	0.0000
0.525	0.0606	0.0000	0.0000	0.0000
0.550	0.2212	0.0000	0.0000	0.0000
0.575	0.4302	0.0000	0.0000	0.0000
0.600	0.6321	0.0000	0.0000	0.0000
0.625	0.7904	0.0000	0.0000	0.0000
0.650	0.8946	0.0000	0.0000	0.0000
0.675	0.9532	0.0000	0.0000	0.0000
0.700	0.9817	0.0000	0.0000	0.0000
0.725	0.9937	0.0000	0.0000	0.0000
0.750	0.9981	0.0000	0.0000	0.0000
0.775	0.9995	0.0000	0.0000	0.0000
0.800	0.9999	0.0000	0.0000	0.0000
0.825	1.0000	0.0000	0.0000	0.0000
0.850	1.0000	0.0000	0.0000	0.0000
0.875	1.0000	0.0000	0.0000	0.0000
0.900	1.0000	0.0000	0.0000	0.0000
0.925	1.0000	0.0000	0.0000	0.0000
0.950	1.0000	0.0000	0.0000	0.0000
0.975	1.0000	0.0000	0.0000	0.0000

Table 8.2: Abbreviated numerical output, ncase=2

```
1.000    1.0000     0.0000      0.0000      0.0000
            .                      .
            .                      .
            .                      .
   Output for t = 0.5, 1, 1.5 removed
            .                      .
            .                      .
            .                      .
```

t = 2.0

r	u1(r,t)	u2(r,t)	u3(r,t)	u4(r,t)
0.000	0.0392	0.7515	0.7515	0.0000
0.025	0.0401	0.7493	0.7493	0.0000
0.050	0.0423	0.7430	0.7430	0.0000
0.075	0.0461	0.7326	0.7326	0.0000
0.100	0.0516	0.7182	0.7182	0.0000
0.125	0.0589	0.6999	0.6999	0.0000
0.150	0.0682	0.6778	0.6778	0.0000
0.175	0.0797	0.6522	0.6522	0.0000
0.200	0.0935	0.6235	0.6235	0.0000
0.225	0.1099	0.5918	0.5918	0.0000
0.250	0.1291	0.5576	0.5576	0.0000
0.275	0.1512	0.5214	0.5214	0.0000
0.300	0.1763	0.4837	0.4837	0.0000
0.325	0.2045	0.4450	0.4450	0.0000
0.350	0.2357	0.4059	0.4059	0.0000
0.375	0.2699	0.3669	0.3669	0.0000
0.400	0.3068	0.3286	0.3286	0.0000
0.425	0.3463	0.2915	0.2915	0.0000
0.450	0.3877	0.2560	0.2560	0.0000
0.475	0.4309	0.2226	0.2226	0.0000
0.500	0.4752	0.1915	0.1915	0.0000
0.525	0.5200	0.1630	0.1630	0.0000
0.550	0.5648	0.1372	0.1372	0.0000
0.575	0.6089	0.1142	0.1142	0.0000
0.600	0.6519	0.0940	0.0940	0.0000
0.625	0.6931	0.0765	0.0765	0.0000
0.650	0.7321	0.0615	0.0615	0.0000
0.675	0.7685	0.0488	0.0488	0.0000
0.700	0.8020	0.0383	0.0383	0.0000
0.725	0.8324	0.0297	0.0297	0.0000
0.750	0.8596	0.0227	0.0227	0.0000
0.775	0.8835	0.0172	0.0172	0.0000
0.800	0.9043	0.0128	0.0128	0.0000
0.825	0.9219	0.0095	0.0095	0.0000
0.850	0.9366	0.0069	0.0069	0.0000

Table 8.2: (*Continued*)

0.875	0.9485	0.0051	0.0051	0.0000
0.900	0.9578	0.0037	0.0037	0.0000
0.925	0.9648	0.0027	0.0027	0.0000
0.950	0.9697	0.0021	0.0021	0.0000
0.975	0.9724	0.0017	0.0017	0.0000
1.000	0.9735	0.0015	0.0015	0.0000

Table 8.2: (*Continued*)

```
[1]  164
[1]  1
 ncase =  3   ncall = 772
 Output for t = 0, 0.5, 1, 1.5 removed

         .                    .
         .                    .
         .                    .

 t = 2.0
       r   u1(r,t)   u2(r,t)   u3(r,t)   u4(r,t)
   0.000    0.0392    0.9375    0.9375    0.0000
   0.025    0.0401    0.9367    0.9367    0.0000
   0.050    0.0423    0.9342    0.9342    0.0000
   0.075    0.0461    0.9301    0.9301    0.0000
   0.100    0.0516    0.9242    0.9242    0.0000
   0.125    0.0589    0.9164    0.9164    0.0000
   0.150    0.0682    0.9066    0.9066    0.0000
   0.175    0.0797    0.8946    0.8946    0.0000
   0.200    0.0935    0.8803    0.8803    0.0000
   0.225    0.1099    0.8635    0.8635    0.0000
   0.250    0.1291    0.8439    0.8439    0.0000
   0.275    0.1512    0.8216    0.8216    0.0000
   0.300    0.1763    0.7962    0.7962    0.0000
   0.325    0.2045    0.7678    0.7678    0.0000
   0.350    0.2357    0.7362    0.7362    0.0000
   0.375    0.2699    0.7016    0.7016    0.0000
   0.400    0.3068    0.6640    0.6640    0.0000
   0.425    0.3463    0.6236    0.6236    0.0000
   0.450    0.3877    0.5808    0.5808    0.0000
   0.475    0.4309    0.5361    0.5361    0.0000
   0.500    0.4752    0.4899    0.4899    0.0000
   0.525    0.5200    0.4430    0.4430    0.0000
   0.550    0.5648    0.3960    0.3960    0.0000
   0.575    0.6089    0.3498    0.3498    0.0000
   0.600    0.6519    0.3050    0.3050    0.0000
```

Table 8.3: Abbreviated numerical output, `ncase=3`

0.625	0.6931	0.2625	0.2625	0.0000
0.650	0.7321	0.2226	0.2226	0.0000
0.675	0.7685	0.1863	0.1863	0.0000
0.700	0.8020	0.1534	0.1534	0.0000
0.725	0.8324	0.1247	0.1247	0.0000
0.750	0.8596	0.0995	0.0995	0.0000
0.775	0.8835	0.0785	0.0785	0.0000
0.800	0.9043	0.0607	0.0607	0.0000
0.825	0.9219	0.0467	0.0467	0.0000
0.850	0.9366	0.0351	0.0351	0.0000
0.875	0.9485	0.0265	0.0265	0.0000
0.900	0.9578	0.0196	0.0196	0.0000
0.925	0.9648	0.0152	0.0152	0.0000
0.950	0.9697	0.0117	0.0117	0.0000
0.975	0.9724	0.0101	0.0101	0.0000
1.000	0.9735	0.0081	0.0081	0.0000

Table 8.3: (*Continued*)

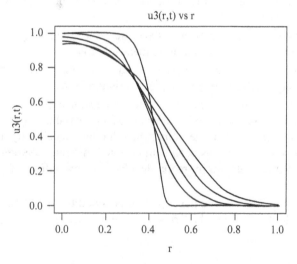

Figure 8.3 $u_3(r, t)$ vs r with $t = 0, 0.5, \ldots, 2$ as a parameter for **ncase=3**

- u_4 remains at the zero IC since no terms have been added to eq. (8.1d) that would move it away from this IC.

This increase in u_3 is also demonstrated in Fig. 8.3 (compared with Fig. 8.2).

- For **ncase=4**, chemotaxis is added to eqs. (8.1a) and (8.1d) through the terms

$$\lambda \frac{\partial}{\partial r} \left(u_1 \frac{\partial u_4}{\partial r} \right), u_{40}[1 - \exp\left(-100(r - r_b)^2\right)] \text{ (in eq. (8.2d))}.$$

```
#
# Diffusion, tumor cell growth, chemotaxis
  if(ncase==4){
      D1=0.01; D2=0.01;
      D3=0.01; D4=0.01;
    alpha1=1;alpha2=1;
     beta1=1; beta2=1;
       lam=1;   u40=1;
  }
```

We can observe the effect of λ in eq. (8.1a) and u_{40} in eq. (8.2d) on the output by moving these parameters away from the zero values of ncase=1,2,3. We can note the following details of this output.

- ncase=4 is confirmed, and the number of calls to pde_1 (Listing 8.2) is ncall = 1107 which is modest.

```
      ncase =  4    ncall = 1107
```

- The IC of eqs. (8.1d), (8.2d) has been modified through the change in u_{40} but is not included in Table 8.4 to conserve space. Also, the output for $t = 0.5, 1, 1.5$ is not included.
- The solution for u_4 is nonzero as expected since the changed IC of eq. (8.1d), that is, eq. (8.2d), affects the subsequent solution.
- u_4 is used in the chemotaxis term of eq. (8.1a) (see pde_1 of Listing 8.2) to provide an increased (more complex) variation in u_1.

This change in u_4 away from zero is also demonstrated in Fig. 8.4.

In summary, the effects of all four dependent variables, u_1, u_2, u_3, u_4, of eqs. (8.1) are now included in the numerical solutions. Specifically, the contribution of chemotaxis to u_1, u_4 has been demonstrated through ncase=4. These various effects can be explored in more detailed by examining the four plots for u_1, u_2, u_3, u_4.

- For ncase=5, CTL-chemokine depletion is added to eqs. (8.1a), (8.1d) through nonzero values for ψ, δ in the terms $-\psi u_1$ and $-\delta u_4$.

```
#
# Diffusion, tumor cell growth, chemotaxis,
# CTL-chemokine depletion
  if(ncase==5){
      D1=0.01; D2=0.01;
      D3=0.01; D4=0.01;
    alpha1=1;alpha2=1;
     beta1=1; beta2=1;
       lam=1;   u40=1;
     delta=1;   psi=1;
  }
```

```
[1]  164
[1]  1
 ncase =  4    ncall = 1107
 Output for t = 0, 0.5, 1, 1.5 removed
                    .                    .
                    .                    .
                    .                    .
 t = 2.0
      r    u1(r,t)    u2(r,t)    u3(r,t)    u4(r,t)
 0.000    1.6874     0.9375     0.9375     0.0392
 0.025    1.6068     0.9367     0.9367     0.0401
 0.050    1.3780     0.9342     0.9342     0.0423
 0.075    1.0646     0.9301     0.9301     0.0461
 0.100    0.7435     0.9242     0.9242     0.0516
 0.125    0.4751     0.9164     0.9164     0.0589
 0.150    0.2868     0.9066     0.9066     0.0682
 0.175    0.1727     0.8946     0.8946     0.0797
 0.200    0.1100     0.8803     0.8803     0.0935
 0.225    0.0766     0.8635     0.8635     0.1099
 0.250    0.0575     0.8439     0.8439     0.1291
 0.275    0.0456     0.8216     0.8216     0.1512
 0.300    0.0373     0.7962     0.7962     0.1763
 0.325    0.0315     0.7678     0.7678     0.2045
 0.350    0.0269     0.7362     0.7362     0.2357
 0.375    0.0237     0.7016     0.7016     0.2699
 0.400    0.0210     0.6640     0.6640     0.3068
 0.425    0.0192     0.6236     0.6236     0.3463
 0.450    0.0175     0.5808     0.5808     0.3877
 0.475    0.0165     0.5361     0.5361     0.4309
 0.500    0.0155     0.4899     0.4899     0.4752
 0.525    0.0150     0.4430     0.4430     0.5200
 0.550    0.0144     0.3960     0.3960     0.5648
 0.575    0.0144     0.3498     0.3498     0.6089
 0.600    0.0142     0.3050     0.3050     0.6519
 0.625    0.0145     0.2625     0.2625     0.6931
 0.650    0.0146     0.2226     0.2226     0.7321
 0.675    0.0153     0.1863     0.1863     0.7685
 0.700    0.0156     0.1534     0.1534     0.8020
 0.725    0.0167     0.1247     0.1247     0.8324
 0.750    0.0173     0.0995     0.0995     0.8596
 0.775    0.0188     0.0785     0.0785     0.8835
 0.800    0.0195     0.0607     0.0607     0.9043
 0.825    0.0214     0.0467     0.0467     0.9219
 0.850    0.0222     0.0351     0.0351     0.9366
```

Table 8.4: Abbreviated numerical output, **ncase=4**

0.875	0.0244	0.0265	0.0265	0.9485
0.900	0.0252	0.0196	0.0196	0.9578
0.925	0.0273	0.0152	0.0152	0.9648
0.950	0.0280	0.0117	0.0117	0.9697
0.975	0.0293	0.0101	0.0101	0.9724
1.000	0.0274	0.0081	0.0081	0.9735

Table 8.4: (*Continued*)

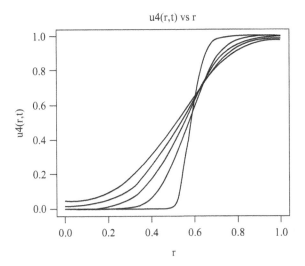

Figure 8.4 $u_4(r, t)$ vs r with $t = 0, 0.5, \ldots, 2$ as a parameter for `ncase=4`

In both equations, the effect is depletion since $-\psi u_1$ and $-\delta u_4$ are negative.
We can note the following details of this output.

– `ncase=5` is confirmed, and the number of calls to `pde_1` (Listing 8.2) is `ncall` = 775 which is modest.

```
ncase =   5    ncall = 775
```

– The output for $t = 0, 0.5, 1, 1.5$ is not included in Table 8.5a to conserve space.
– The solution for u_1 is complicated as a result of the combined effect of diffusion, tumor cell growth, chemotaxis, and for `ncase=5`, CTL-chemokine depletion.
– The solutions for u_2 and u_3 remain the same since to this point, the parameter values have not been selected to differentiate these solutions.

The complexity of the u_1 solution is demonstrated in Fig. 8.5a. Also, the solution curves are not as completely smooth as we might expect, suggesting a gridding effect in r. This can be studied by repeating the solutions for a larger number of grid points. For example, `nr=81` (changed in Listing 8.1) gives the following abbreviated output (numerical in Table 8.5b and graphical in Fig. 8.5b)

```
[1]  164
[1]  1
 ncase =  5    ncall = 775
 Output for t = 0, 0.5, 1, 1.5 removed

            .                    .
            .                    .
            .                    .

 t = 2.0
       r    u1(r,t)    u2(r,t)    u3(r,t)    u4(r,t)
  0.000    0.2253     0.9375     0.9375     0.0053
  0.025    0.2241     0.9367     0.9367     0.0054
  0.050    0.2196     0.9342     0.9342     0.0057
  0.075    0.2121     0.9301     0.9301     0.0062
  0.100    0.2010     0.9242     0.9242     0.0070
  0.125    0.1865     0.9164     0.9164     0.0080
  0.150    0.1687     0.9066     0.9066     0.0092
  0.175    0.1485     0.8946     0.8946     0.0108
  0.200    0.1267     0.8803     0.8803     0.0127
  0.225    0.1050     0.8635     0.8635     0.0149
  0.250    0.0845     0.8439     0.8439     0.0175
  0.275    0.0666     0.8216     0.8216     0.0205
  0.300    0.0518     0.7962     0.7962     0.0239
  0.325    0.0405     0.7678     0.7678     0.0277
  0.350    0.0321     0.7362     0.7362     0.0319
  0.375    0.0263     0.7016     0.7016     0.0365
  0.400    0.0224     0.6640     0.6640     0.0415
  0.425    0.0198     0.6236     0.6236     0.0469
  0.450    0.0182     0.5808     0.5808     0.0525
  0.475    0.0172     0.5361     0.5361     0.0583
  0.500    0.0166     0.4899     0.4899     0.0643
  0.525    0.0163     0.4430     0.4430     0.0704
  0.550    0.0162     0.3960     0.3960     0.0764
  0.575    0.0164     0.3498     0.3498     0.0824
  0.600    0.0168     0.3050     0.3050     0.0882
  0.625    0.0173     0.2625     0.2625     0.0938
  0.650    0.0181     0.2226     0.2226     0.0991
  0.675    0.0190     0.1863     0.1863     0.1040
  0.700    0.0201     0.1534     0.1534     0.1085
  0.725    0.0214     0.1247     0.1247     0.1127
  0.750    0.0228     0.0995     0.0995     0.1163
  0.775    0.0244     0.0785     0.0785     0.1196
  0.800    0.0261     0.0607     0.0607     0.1224
  0.825    0.0279     0.0467     0.0467     0.1248
  0.850    0.0297     0.0351     0.0351     0.1267
```

Table 8.5a: Abbreviated numerical output, ncase=5

0.875	0.0314	0.0265	0.0265	0.1284
0.900	0.0331	0.0196	0.0196	0.1296
0.925	0.0344	0.0152	0.0152	0.1306
0.950	0.0356	0.0117	0.0117	0.1312
0.975	0.0362	0.0101	0.0101	0.1316
1.000	0.0368	0.0081	0.0081	0.1317

Table 8.5a: (*Continued*)

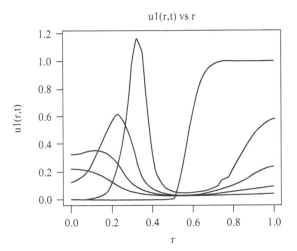

Figure 8.5a $u_1(r, t)$ vs r with $t = 0, 0.5, \ldots, 2$ as a parameter for `ncase=5`

The numerical solution in Table 8.5b appears to be smoother than in Table 8.5a. This is confirmed in Fig. 8.5b when compared with Fig. 8.5a. In other words, the gridding with `nr=41` is too coarse and produces numerical artifacts.The increase in `ncall` from 775 to 2290 results from the increase in `nr` from 41 to 81.

In summary, the solutions for u_1, u_2, u_3, u_4 are becoming increasingly complex as additional effects and nonzero terms are added to eqs. (8.1). The step-by-step procedure (through `ncase=1,2,3,4,5`) identifies the contribution of each nonzero term added to eqs. (8.1). The solution with `nr=81` indicates that a larger number of grid points should be retained for the subsequent solutions.

- For `ncase=6`, CTL-chemokine proliferation is added to eqs. (8.1a), (8.1d) through nonzero values for γ_1, γ_2 in the terms $\dfrac{\gamma_1 u_1 u_3}{\eta_1 + u_3}$ and $\dfrac{\gamma_2 u_1 u_3}{\eta_1 + u_3}$.

```
#
# Diffusion, tumor cell growth, chemotaxis,
# CTL-chemokine depletion, CTL proliferation
  if(ncase==6){
    D1=0.01; D2=0.01;
    D3=0.01; D4=0.01;
```

```
[1] 324
[1] 1
 ncase =  5    ncall = 2290
 Output for t = 0, 0.5, 1, 1.5 removed
                .                    .
                .                    .
                .                    .
                .                    .
  t = 2.0
     r    u1(r,t)    u2(r,t)    u3(r,t)    u4(r,t)
 0.000    0.2256     0.9375     0.9375     0.0053
 0.012    0.2253     0.9373     0.9373     0.0053
 0.025    0.2242     0.9367     0.9367     0.0054
 0.038    0.2224     0.9357     0.9357     0.0055
 0.050    0.2198     0.9342     0.9342     0.0057
 0.062    0.2164     0.9324     0.9324     0.0060
 0.075    0.2121     0.9301     0.9301     0.0062
 0.088    0.2071     0.9273     0.9273     0.0066
 0.100    0.2011     0.9242     0.9242     0.0070
                .                    .
                .                    .
                .                    .
                .                    .

  Output for r = 0.112 to 0.888 removed

                .                    .
                .                    .
                .                    .
 0.900    0.0330     0.0197     0.0197     0.1296
 0.913    0.0337     0.0171     0.0171     0.1301
 0.925    0.0344     0.0150     0.0150     0.1306
 0.938    0.0350     0.0132     0.0132     0.1309
 0.950    0.0355     0.0118     0.0118     0.1312
 0.963    0.0358     0.0108     0.0108     0.1315
 0.975    0.0361     0.0100     0.0100     0.1316
 0.988    0.0363     0.0096     0.0096     0.1317
 1.000    0.0364     0.0094     0.0094     0.1317
```

Table 8.5b: Abbreviated numerical output, ncase=5, nr=81

```
        alpha1=1;alpha2=1;
         beta1=1; beta2=1;
            lam=1;   u40=1;
          delta=1;   psi=1;
           gam1=1;  gam2=1;
      }
```

Abbreviated output follows (numerical in Table 8.6 and graphical in Fig. 8.6).

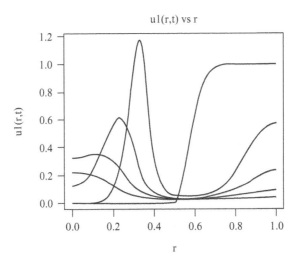

Figure 8.5b $u_1(r, t)$ vs r with $t = 0, 0.5, \ldots, 2$ as a parameter for `ncase=5`, `nr=81`

We can note the following details of this output.

- `ncase=6` is confirmed, and the number of calls to pde_1 (Listing 8.2) is `ncall = 2610`.

      ncase = 6   ncall = 2610

- The output for $t = 0, 0.5, 1, 1.5$ is not included in Table 8.6 to conserve space.
- The solution for u_1 is complicated as a result of the combined effect of diffusion, tumor cell growth, chemotaxis, CTL-chemokine depletion and for `ncase=6` CTL-chemokine proliferation.
- The solutions for u_2 and u_3 remain the same since to this point, the parameter values have not been selected to differentiate these solutions.

    The complexity of the u_1 solution is demonstrated in Fig. 8.6. The solutions are smooth for `nr=81`.

- For `ncase=7`, virus replication is added to eqs. (8.1b), (8.1c) through nonzero values for θ_1, θ_2 in the terms $-\dfrac{\theta_1 u_2 u_3}{\eta_2 + u_3}$ and $+\dfrac{\theta_2 u_2 u_3}{\eta_2 + u_3}$.

```
#
# Diffusion, tumor cell growth, chemotaxis,
# CTL-chemokine depletion, CTL proliferation,
# virus replication
  if(ncase==7){
      D1=0.01; D2=0.01;
      D3=0.01; D4=0.01;
    alpha1=1;alpha2=1;
    beta1=1; beta2=1;
```

```
[1]  324
[1]  1
 ncase =  6    ncall = 2610
 Output for t = 0, 0.5, 1, 1.5 removed
                .                    .
                .                    .
                .                    .

 t = 2.0
      r    u1(r,t)    u2(r,t)    u3(r,t)    u4(r,t)
 0.000     0.1815     0.9375     0.9375     0.1166
 0.012     0.1815     0.9373     0.9373     0.1165
 0.025     0.1814     0.9367     0.9367     0.1165
 0.038     0.1814     0.9357     0.9357     0.1164
 0.050     0.1813     0.9342     0.9342     0.1162
 0.062     0.1812     0.9324     0.9324     0.1160
 0.075     0.1810     0.9301     0.9301     0.1157
 0.088     0.1808     0.9273     0.9273     0.1154
 0.100     0.1806     0.9242     0.9242     0.1151
                .                    .
                .                    .
                .                    .

 Output for r = 0.112 to 0.888 removed
                .                    .
                .                    .
                .                    .

 0.900     0.0346     0.0197     0.0197     0.1302
 0.913     0.0353     0.0171     0.0171     0.1306
 0.925     0.0359     0.0150     0.0150     0.1310
 0.938     0.0364     0.0132     0.0132     0.1313
 0.950     0.0369     0.0118     0.0118     0.1316
 0.963     0.0372     0.0108     0.0108     0.1318
 0.975     0.0375     0.0100     0.0100     0.1319
 0.988     0.0377     0.0096     0.0096     0.1320
 1.000     0.0377     0.0094     0.0094     0.1320
```

Table 8.6: Abbreviated numerical output, ncase=6

```
        lam=1;    u40=1;
      delta=1;    psi=1;
       gam1=1;   gam2=1;
     theta1=1;theta2=1;
   }
```

Abbreviated output follows (numerical in Table 8.7 and graphical in Figs. 8.7a, 8.7b).

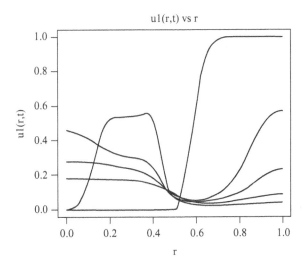

Figure 8.6 $u_1(r, t)$ vs r with $t = 0, 0.5, \ldots, 2$ as a parameter for `ncase=6`, `nr=81`

We can note the following details of this output.

– `ncase=7` is confirmed, and the number of calls to `pde_1` (Listing 8.2) is `ncall = 2788`.

```
ncase =  7    ncall = 2788
```

– The output for $t = 0, 0.5, 1, 1.5$ is not included in Table 8.7 to conserve space.
– The solutions for u_2 and u_3 are now substantially different, which is to be expected since the terms in θ_1, θ_2 in eqs. (8.1b), (8.1c) are of opposite sign. The differences are indicated in Figs. 8.7a and 8.7b.
• For `ncase=8`, immune cell supply is added to eq. (8.1a), through a nonzero value for σ in the term $\sigma h(r)$.

```
#
# Diffusion, tumor cell growth, chemotaxis,
# CTL-chemokine depletion, CTL proliferation,
# virus replication, immune cell supply
  if(ncase==8){
      D1=0.01; D2=0.01;
      D3=0.01; D4=0.01;
    alpha1=1;alpha2=1;
     beta1=1; beta2=1;
       lam=1;    u40=1;
     delta=1;    psi=1;
      gam1=1;   gam2=1;
    theta1=1;theta2=1;
     sigma=1;
  }
```

```
[1] 324
[1] 1
 ncase =  7   ncall = 2788
 Output for t = 0, 0.5, 1, 1.5 removed
             .                    .
             .                    .
             .                    .

 t = 2.0
     r    u1(r,t)    u2(r,t)    u3(r,t)    u4(r,t)
 0.000    0.1872     0.5416     1.1956     0.1314
 0.012    0.1872     0.5415     1.1954     0.1314
 0.025    0.1871     0.5411     1.1947     0.1313
 0.038    0.1871     0.5406     1.1935     0.1312
 0.050    0.1870     0.5398     1.1918     0.1310
 0.062    0.1869     0.5387     1.1897     0.1308
 0.075    0.1868     0.5375     1.1870     0.1305
 0.088    0.1866     0.5359     1.1838     0.1302
 0.100    0.1865     0.5342     1.1801     0.1298
             .                    .
             .                    .
             .                    .

 Output for r = 0.112 to 0.888 removed
             .                    .
             .                    .
             .                    .
 0.900    0.0348     0.0162     0.0230     0.1303
 0.913    0.0354     0.0142     0.0199     0.1307
 0.925    0.0360     0.0125     0.0173     0.1311
 0.938    0.0366     0.0112     0.0152     0.1314
 0.950    0.0370     0.0100     0.0135     0.1316
 0.963    0.0374     0.0092     0.0123     0.1318
 0.975    0.0377     0.0086     0.0114     0.1320
 0.988    0.0378     0.0083     0.0109     0.1320
 1.000    0.0379     0.0081     0.0106     0.1321
```

Table 8.7: Abbreviated numerical output, ncase=7

Abbreviated output follows (numerical in Tables 8.8a 8.8b and graphical in Figs. 8.8a, 8.8b).

We can note the following details of this output.

– ncase=8 is confirmed, and the number of calls to pde_1 (Listing 8.2) is ncall = 4110.

```
 ncase =  8   ncall = 4110
```

– The output for $t = 0, 0.5, 1, 1.5$ is not included in Table 8.8a to conserve space.

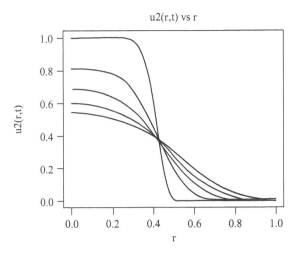

Figurè 8.7a $u_2(r, t)$ vs r with $t = 0, 0.5, \ldots, 2$ as a parameter for `ncase=7`

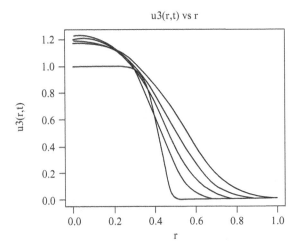

Figure 8.7b $u_3(r, t)$ vs r with $t = 0, 0.5, \ldots, 2$ as a parameter for `ncase=7`

- The solution for u_1 has a numerical oscillation as reflected in the numerical
  output for u_1 in Table 8.8a and in Fig. 8.8a. The cause of this oscillation is
  explained next.

  The oscillation in Fig. 8.8a results from the discontinuity in the term $\sigma h(r)$
  in eq. (8.1a) (recall the unit step in `h(r)` in Listing 8.3). More generally, finite
  difference (FD) approximations such as in `dss004` will oscillate in response to
  discontinuities. The oscillations could be more pronounced without the smoothing
  effect of the diffusion expressed in terms of the diffusivity D_1 in eq. (8.1a).

```
[1] 324
[1] 1
 ncase =  8    ncall = 4110
 Output for t = 0, 0.5, 1, 1.5 removed
              .                    .
              .                    .
              .                    .
              .                    .
 t = 2.0
      r    u1(r,t)   u2(r,t)   u3(r,t)   u4(r,t)
 0.000     0.4165    0.5416    1.1956    0.2420
 0.012     0.3856    0.5415    1.1954    0.2349
 0.025     0.3895    0.5411    1.1947    0.2358
 0.038     0.3824    0.5406    1.1935    0.2340
 0.050     0.3893    0.5398    1.1918    0.2355
 0.062     0.3825    0.5387    1.1897    0.2338
 0.075     0.3897    0.5375    1.1870    0.2352
 0.088     0.3829    0.5359    1.1838    0.2334
 0.100     0.3902    0.5342    1.1801    0.2347
              .                    .
              .                    .
              .                    .
 Output for r = 0.112 to 0.888 removed
              .                    .
              .                    .
              .                    .
 0.900     0.9016    0.0162    0.0230    0.1371
 0.913     0.8470    0.0142    0.0199    0.1362
 0.925     0.8387    0.0125    0.0173    0.1360
 0.938     0.7952    0.0112    0.0152    0.1354
 0.950     0.7984    0.0100    0.0135    0.1354
 0.963     0.7648    0.0092    0.0123    0.1349
 0.975     0.7785    0.0086    0.0114    0.1350
 0.988     0.7621    0.0083    0.0109    0.1348
 1.000     0.8555    0.0081    0.0106    0.1359
```

Table 8.8a: Abbreviated numerical output, ncase=8

To circumvent this numerical error, a smoother function than $h(r)$, such as a unit ramp as implemented in the function ramp in Listing 8.4, can be used (in Listing 8.2).

```
ramp=function(r){
#
# Function ramp defines a unit ramp function
# for the supply of immune cells
```

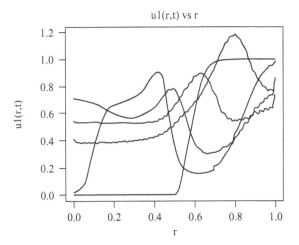

u1(r,t) vs r

Figure 8.8a $u_1(r, t)$ vs r with $t = 0, 0.5, \ldots, 2$ as a parameter for `ncase=8`

```
#
  rb=0.5;
  if(r<rb){
    ramp=0;
  }else if((r>=rb)&(r<=1.1*rb)){
    ramp=(r-rb)/(0.1*rb);
  }else{
    ramp=1;
  }
#
# Return function
  return(c(ramp));
}
```

Listing 8.4: `ramp` for a unit ramp

We can note the following details of Listing 8.4.

– The function is defined.

```
    ramp=function(r){
  #
  # Function ramp defines a unit ramp function
  # for the supply of immune cells
```

The details of the ramp follow.

■ For $r < r_b$ (r_b is used in ICs (8.2)), the function is zero.

```
  #
    rb=0.5;
```

```
if(r<rb){
  ramp=0;
```

■ For $r_b \leq r \leq 1.1r_b$, the function is a linear increase in r, that is, a linear function for $0.5 \leq r \leq 0.550$ with slope $1/0.1r_b$.

```
}else if((r>=rb)&(r<=1.1*rb)){
  ramp=(r-rb)/(0.1*rb);
```

■ For $r > 1.1r_b$, the function is one.

```
}else{
  ramp=1;
}
```

$ramp(r)$ is an approximation to the unit step $h(r)$ in eq. (8.1a), but since it has a finite slope, it does not produce the oscillations in the solution of eq. (8.1a) as reflected in Fig. 8.8a. Rather, the solution of Fig. 8.8b is smooth.

Abbreviated output is given in Table 8.8b and Fig. 8.8b.

The solution for u_1 in Table 8.8b is noticeably smoother than in Table 8.8a, e.g., near $r = 1$.

The solution in Fig. 8.8b is clearly smoother than in Fig. 8.8a.

To use ramp in place of h, the following changes were made.

– In Listing 8.1

```
source("pde_1.R") ;source("h.R");
out=ode(func=pde_1,times=tout,y=u0);
```

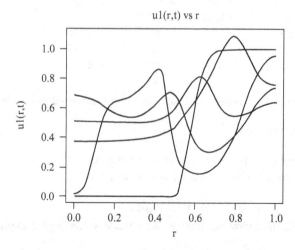

Figure 8.8b $u_1(r,t)$ vs r with $t = 0, 0.5, \ldots, 2$ as a parameter for **ncase=8** and a ramp

```
[1] 324
[1] 1
 ncase =  8    ncall = 4200
 Output for t = 0, 0.5, 1, 1.5 removed
                .                    .
                .                    .
                .                    .
  t = 2.0
       r    u1(r,t)     u2(r,t)     u3(r,t)     u4(r,t)
  0.000     0.3677      0.5416      1.1956      0.2245
  0.012     0.3684      0.5415      1.1954      0.2248
  0.025     0.3684      0.5411      1.1947      0.2247
  0.038     0.3686      0.5406      1.1935      0.2247
  0.050     0.3685      0.5398      1.1918      0.2245
  0.062     0.3688      0.5387      1.1897      0.2244
  0.075     0.3688      0.5375      1.1870      0.2242
  0.088     0.3691      0.5359      1.1838      0.2240
  0.100     0.3691      0.5342      1.1801      0.2237
                .                    .
                .                    .
                .                    .
    Output for r = 0.112 to 0.888 removed
                .                    .
                .                    .
                .                    .
  0.900     0.8549      0.0162      0.0230      0.1367
  0.913     0.8287      0.0142      0.0199      0.1361
  0.925     0.8071      0.0125      0.0173      0.1357
  0.938     0.7897      0.0112      0.0152      0.1354
  0.950     0.7764      0.0100      0.0135      0.1352
  0.963     0.7665      0.0092      0.0123      0.1350
  0.975     0.7599      0.0086      0.0114      0.1349
  0.988     0.7560      0.0083      0.0109      0.1349
  1.000     0.7552      0.0081      0.0106      0.1348
```

Table 8.8b: Abbreviated numerical output, ncase=8 and a ramp

was replaced with

```
source("pde_1.R") ;source("ramp.R");
out=ode(func=pde_1a,times=tout,y=u0);
```

Note the use of pde_1a in place of pde_1.

– In pde_1, the following changes were made to Listing 8.2 to provide pde_1a.

```
pde_1=function(t,u,parms) {
lam*uctr[i]+sigma*h(r[i])-psi*u1[i]+
```

were changed

```
pde_1a=function(t,u,parms) {
lam*uctr[i]+sigma*ramp(r[i])-psi*u1[i]+
(in two places)
```

Otherwise, the code with $h(r)$ and $ramp(r)$ in eq. (8.1a) is the same.

In summary, MOL solutions are sensitive to the smoothness properties of the terms in the PDEs (and ICs, BCs). PDEs with discontinuities are generally termed *Riemann problems*, and they usually require some special treatment, such as the use of a ramp or other relatively smooth function in place of the discontinuity.

- For ncase=9, tumor cell death is added to eqs. (8.1a), (8.1c) through nonzero values for ν, μ in the terms $-\nu u_1 u_3$ and $-\mu u_1 u_3$. These terms are negative and therefore reduce u_1 and most importantly, u_3. Thus, in eq. (8.1c), $-\mu u_1 u_3$ reflects the reduction in tumor cells (u_3) from the effect of CTL-chemokine (u_1).

```
#
# Diffusion, tumor cell growth, chemotaxis,
# CTL-chemokine depletion, CTL proliferation,
# virus replication, immune cell supply,
# tumor cell death
  if(ncase==9){
     D1=0.01; D2=0.01;
     D3=0.01; D4=0.01;
    alpha1=1;alpha2=1;
     beta1=1; beta2=1;
       lam=1;   u40=1;
     delta=1;   psi=1;
      gam1=1;  gam2=1;
    theta1=1;theta2=1;
     sigma=1;
        nu=1;    mu=1;
  }
```

Abbreviated output follows (numerical in Table 8.9 and graphical in Figs. 8.9a, 8.9b). These numerical results were produced with the inclusion of ramp in place of h in eq. (8.1a) as discussed for ncase = 8.

We can note the following details of this output.

- ncase=9 is confirmed, and the number of calls to pde_1 (Listing 8.2) is ncall = 4161.

```
ncase =  9   ncall = 4161
```

- The output for $t = 0, 0.5, 1, 1.5$ is not included in Table 8.9 to conserve space.
- The solutions for u_1 and u_3 reflect the reduction in tumor cell concentration, u_3 from eq. (8.1c). For example, from Tables 8.8 and 8.9,

```
[1]  324
[1]  1
 ncase =  9    ncall = 4161
 Output for t = 0, 0.5, 1, 1.5 removed
                .                         .
                .                         .
                .                         .

 t = 2.0
       r    u1(r,t)     u2(r,t)     u3(r,t)     u4(r,t)
   0.000     0.1544      0.5759      0.9674      0.1190
   0.012     0.1547      0.5758      0.9669      0.1191
   0.025     0.1549      0.5755      0.9659      0.1192
   0.038     0.1553      0.5751      0.9641      0.1193
   0.050     0.1556      0.5744      0.9617      0.1195
   0.062     0.1563      0.5736      0.9585      0.1197
   0.075     0.1569      0.5725      0.9547      0.1200
   0.088     0.1578      0.5713      0.9501      0.1204
   0.100     0.1587      0.5698      0.9447      0.1207
                .                         .
                .                         .
                .                         .

   Output for r = 0.112 to 0.888 removed
                .                         .
                .                         .
                .                         .
   0.900     0.6873      0.0169      0.0097      0.1328
   0.913     0.6859      0.0147      0.0083      0.1329
   0.925     0.6855      0.0129      0.0072      0.1330
   0.938     0.6851      0.0115      0.0063      0.1331
   0.950     0.6853      0.0103      0.0056      0.1331
   0.963     0.6853      0.0095      0.0051      0.1332
   0.975     0.6858      0.0088      0.0047      0.1332
   0.988     0.6858      0.0085      0.0045      0.1332
   1.000     0.6867      0.0083      0.0044      0.1332
```

Table 8.9: Abbreviated numerical output, ncase=9

```
        r = 0
            r    u1(r,t)     u2(r,t)     u3(r,t)     u4(r,t)
        Table 8.8
        0.000      0.3677      0.5416      1.1956      0.2245
        Table 8.9
        0.000      0.1544      0.5759      0.9674      0.1190
          r = 1
            r    u1(r,t)     u2(r,t)     u3(r,t)     u4(r,t)
        Table 8.8
```

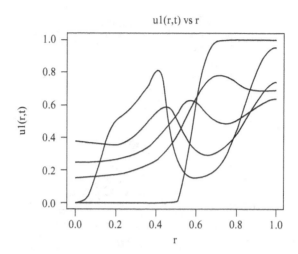

Figure 8.9a $u_1(r,t)$ vs r with $t = 0, 0.5, \ldots, 2$ as a parameter for `ncase=9`

```
1.000     0.7552     0.0081     0.0106     0.1348
Table 8.9
1.000     0.6867     0.0083     0.0044     0.1332
```

For `r = 1`, $u_3(r = 1, t = 2)$ has been reduced from 0.0106 to 0.0044 (infected cells) while $u_2(r = 1, t = 2)$ has no significant reduction, 0.0081, 0.0083 (uninfected cells). μ, ν in eqs. (8.1a), (8.1c) are clearly key parameters in the reduction of infected cancer cells and can be studied in detail using Listings 8.1, 8.2 and 8.4 (with `ramp` in place of h). For example, $-\nu u_1 u_3$ and $-\mu u_1 u_3$ could be computed from the numerical solutions, u_1, u_3 and plotted against r with t as a parameter.

Fig. 8.9a indicates that the complex response of u_1 remains smooth (as discussed for ncase = 8). Fig. 8.9b also indicates a complex response for u_3 that could be studied in detail by varying, for example, ν, μ.

(8.4) Summary and conclusions

A methodology for the MOL implementation of the 1D 4-PDE model of eqs. (8.1), (8.2) and (8.3) is discussed in this chapter. In particular, the step-by-step procedure of adding terms to the RHS of the PDEs offers a method for systematically examining the contribution of the terms as they are added. This has the advantage that if during the procedure, a failure in the model execution, or an obvious numerical distortion (e.g., oscillation from coarse gridding or a discontinuity) is revealed, steps can be taken to correct the problem (e.g., increasing the number of grid points, replacement of the discontinuity with a smoother function).

The preceding example also illustrates an essential evaluation of the numerical solution as an error analysis. This was demonstrated with h-refinement by increasing nr=41

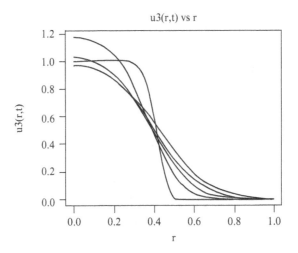

Figure 8.9b $u_3(r, t)$ vs r with $t = 0, 0.5, \ldots, 2$ as a parameter for `ncase=9`

to `nr=81` (the grid spacing in a FD approximation is usually denoted with h, and there-fore the procedure of changing the spacing is termed h-refinement).

Another possibility for an error analysis of the numerical solution is p-refinement in which the order of the (FD) approximations is changed. For example, `dss004` in Listing 8.2 based on fourth-order FDs so that $p = 4$ could be replaced by `dss006` based on sixth-order approximations so that $p = 6$. The two solutions could then be compared as a check on the spatial convergence of the solution.

More generally, this discussion indicates that some experimentation in developing a MOL solution to a PDE system is usually required, with attention to the details of the solution as the analysis proceeds. In other words, MOL analysis is not a mechanical procedure with a guaranteed final satisfactory result. Rather, each problem should be considered as a new problem with success in achieving a solution of acceptable accuracy dependent on careful programming and evaluation of the solution.

Reference

[1] Malinzi, J., P. Sibanda, and H. Mambili-Mamboundou (2015), Analysis of virotherapy in solid tumor invasion, *Math. Biosci.*, **263**, 102–110; an extensive list of references is included with this reference

9

TUMOR CELL DENSITY
IN GLIOBLASTOMAS

This chapter pertains to a mathematical model for the distribution of cancer cells in glioblastomas, an aggressive form of cancer of the brain [1]. The 1D, 1-PDE model is based on diffusion of the cells in white and gray tissue, and a growth law of varying form, that is, a diffusion-reaction PDE [2,3].

The intent of this chapter is to:

- Present a 1-PDE model for cell density dynamics in glioblastomas including the required initial conditions (ICs) and boundary conditions (BCs).

- Discuss the format of the model as a variable coefficient, diffusion-reaction (or parabolic) PDE in 1D.

- Present a method of lines (MOL) solution, including the use of library routines for integration of the PDE derivatives in time and space.

- Present the computed model solution in numerical and graphical (plotted) format.

- Investigate three forms of the cancer cell volumetric growth rate to determine their specific effect on the solutions.

- Discuss the performance of the algorithms used to compute the solution. In particular, h-refinement is used to infer the accuracy of the approximation of the spatial derivatives that model diffusion when the diffusivity changes discontinuously between white and gray matter.

A modification of the 1-PDE model reported in [3] is presented next, followed by a MOL solution programmed in R.

Method of Lines PDE Analysis in Biomedical Science and Engineering, First Edition. William E. Schiesser.
© 2016 John Wiley & Sons, Inc. Published 2016 by John Wiley & Sons, Inc.
Companion website: www.wiley.com/go/Schiesser/PDE_Analysis

(9.1) 1D PDE model

The PDE originally presented in 1D Cartesian coordinates is expressed in 1D spherical coordinates to better reflect the geometry of a tumor[1]

$$\frac{\partial u}{\partial t} = \frac{1}{r^2}\frac{\partial}{\partial r}\left(r^2 D(r)\frac{\partial u}{\partial r}\right) + f(u)$$

or

$$\frac{\partial u}{\partial t} = D(r)\frac{\partial^2 u}{\partial r^2} + \frac{dD(r)}{dr}\frac{\partial u}{\partial r} + D(r)\frac{2}{r}\frac{\partial u}{\partial r} + f(u) \tag{9.1}$$

where

u	tumor cell density
r	radial coordinate in the tumor
t	time
$D(r)$	cell diffusivity as a function of r
$f(u)$	inhomogeneous, source term for cell volumetric growth rate

In eq. (9.1), the diffusion term $\dfrac{1}{r^2}\dfrac{\partial}{\partial r}\left(r^2 D(r)\dfrac{\partial u}{\partial r}\right)$ is expanded to facilitate the numerical solution as explained subsequently.

Eq. (9.1) is first order in t and second order in r. It therefore requires one IC and two BCs.

$$u(r, t = 0) = g(r); \quad \frac{\partial u(r = 0, t)}{\partial r} = \frac{\partial u(r = r_0, t)}{\partial r} = 0 \tag{9.2a,b,c}$$

where $g(r)$ is a function to be specified. Eqs. (9.2b,c) are homogeneous (zero) Neumann BCs for symmetry at $r = 0$ and no diffusion of cells across the tumor boundary $r = r_0$, respectively.

Three cases for $f(u)$ in eq. (9.1) are considered.

Case $f(u)$

ncase = 1: $\rho u(r, t)$

linear

ncase = 2: $\rho u(r, t)\left(1 - \dfrac{u(r, t)}{u_{max}}\right)$ $\tag{9.3}$

logistic

ncase = 3: $-\rho u(r, t)\ln\left(\dfrac{u(r, t)}{e^{k/d}}\right)$

Gompertz

---

[1]In accordance with the usual notation for PDEs in the numerical analysis literature, u is used to denote the PDE dependent variable.

where [3],

Parameter	Interpretation
ρ	cell proliferation rate
u_{max}	carrying capacity of tissue
k	tumor growth rate
d	tumor decay rate

These three cases are programmed and discussed in the routines that follow.

(9.2) MOL routines

The main program and subordinate MOL/ODE routine are discussed next.

(9.2.1) Main program

The main program for eqs. (9.1), (9.2), and (9.3) is in Listing 9.1.

```
#
# Tumor cell density in glioblastomas
#
# Delete previous workspaces
  rm(list=ls(all=TRUE))
#
# Access ODE integrator
  library("deSolve");
#
# Access functions for numerical solutions
  setwd("g:/chap9");
  source("pde_1.R");
  source("dss004.R");
  source("dss044.R");
#
# Level of output
#
#   ip = 1 - graphical (plotted) solutions
#            (u(r,t)) only
#
#   ip = 2 - numerical and graphical output
#
  ip=2;
#
# Select case
  ncase=1;
#
```

```
# Parameters
  rho=0.012;umax=62.5;eps=0.01;k=1;d=1;
#
# Grid in r
  nr=101;r0=25;
  r=seq(from=0,to=r0,by=(r0-0)/(nr-1));
#
# Diffusivity
  D=rep(0,nr);
  for(i in 1:nr){
    if(i<=16){D[i]=0.13;}
    if((i>16)&(i<=86)){D[i]=0.65;}
    if(i> 86){D[i]=0.13;}
  }
#
# Display D(r)
  if(ip==2){
    for(i in 1:nr){
      cat(sprintf("\n i = %3d   r = %4.2f   D(r) = %5.3f",
                   i,r[i],D[i]));
    }
  }
#
# dD/dr
  dDdr=rep(0,nr);
  for(i in 1:nr){
    if(i< 16){dDdr[i]=0;}
    if(i==16){dDdr[i]=(D[i+1]-D[i-1])/0.5;}
    if((i>16)&(i<86)){dDdr[i]=0;}
    if(i==86){dDdr[i]=(D[i+1]-D[i-1])/0.5;}
    if(i> 86){dDdr[i]=0;}
  }
#
# Display dD/dr
  if(ip==2){
    for(i in 1:nr){
      cat(sprintf("\n i = %3d   r = %4.2f   dD/dr = %5.3f",
                   i,r[i],dDdr[i]));
    }
  }
#
# Initial conditions
  u0=rep(0,nr);
  fact=1/((2*pi)^0.5*eps);
  for(i in 1:nr){
      u0[i]=exp(-0.5*(r[i]-12.5)^2/eps);
```

```
    }
  u0=fact*u0;
  ncall=0;
#
# Write selected parameters
  cat(sprintf("\n\n ncase = %2d",ncase));
#
# Write heading
  if(ip==1){
    cat(sprintf("\n Graphical output only\n"));
  }
#
# Independent variable for ODE integration
  nout=6;t0=0;tf=15;
  tout=seq(from=t0,to=tf,by=(tf-t0)/(nout-1));
#
# ODE integration
  out=lsodes(y=u0,times=tout,func=pde_1);
  nrow(out)
  ncol(out)
#
# Arrays for plotting numerical solution
  u_plot1=matrix(0,nrow=nr,ncol=nout);
  u_plot2=matrix(0,nrow=nr,ncol=nout-1);
  for(it in 1:nout){
    for(i in 1:nr){
      u_plot1[i,it]=out[it,i+1];
      if(it>1){u_plot2[i,it-1]=u_plot1[i,it];}
    }
  }
#
# Display numerical solution
  if(ip==2){
    for(it in 1:nout){
      cat(sprintf(
      "\n    t        r     u(r,t)\n"));
      for(i in 1:nr){
        cat(sprintf("%5.1f%8.2f%10.3f\n",
        tout[it],r[i],u_plot1[i,it]));
      }
    }
  }
#
# Calls to ODE routine
  cat(sprintf("\n\n ncall = %5d\n\n",ncall));
#
```

```
# Plot u with t = 0
  par(mfrow=c(1,1));
  matplot(x=r,y=u_plot1,type="l",xlab="r",
          ylab="u(r,t), t=0,3,...,15",xlim=c(0,r0),lty=1,
          main="u(r,t); t=0,3,...,15;",lwd=2,col="black");
#
# Plot u without t = 0
  par(mfrow=c(1,1));
  matplot(x=r,y=u_plot2,type="l",xlab="r",
          ylab="u(r,t), t=3,...,15",xlim=c(0,r0),lty=1,
          main="u(r,t); t=3,...,15;",lwd=2,col="black");
```

Listing 9.1: Main program for eqs. (9.1) to (9.3)

We can note the following details about Listing 9.1.

- Previous workspaces are cleared.

```
  #
  # Tumor cell density in glioblastomas
  #
  # Delete previous workspaces
    rm(list=ls(all=TRUE))
```

- The R library of ODE integrators, deSolve, and the routines for the MOL solution of eqs. (9.1) to (9.3) are accessed. The setwd (set working directory) will be edited for the local computer. Note also that / is used rather than the usual \.

```
  #
  # Access ODE integrator
    library("deSolve");
  #
  # Access functions for numerical solutions
    setwd("g:/chap9");
    source("pde_1.R");
    source("dss004.R");
    source("dss044.R");
```

- A level of numerical output is selected.

```
  #
  # Level of output
  #
  #   ip = 1 - graphical (plotted) solutions
  #           (u(r,t)) only
  #
```

```
#    ip = 2 - numerical and graphical output
#
   ip=2;
```

- One of the cases in eqs. (9.3) is selected

```
#
# Select case
   ncase=1;
```

- The parameters in eq. (9.1) are defined numerically. These values are taken from [3] (rho,umax,eps) or assigned to complete the coding (k,d).

```
#
# Parameters
   rho=0.012;umax=62.5;eps=0.01;k=1;d=1;
```

- A grid in r with 101 points is defined for the interval $0 \le r \le r_0$ with the seq utility. The radius of the spherical tumor is $r_0 = 25$ mm and the grid spacing is therefore $25/(101 - 1) = 0.25$ mm.

```
#
# Grid in r
   nr=101;r0=25;
   r=seq(from=0,to=r0,by=(r0-0)/(nr-1));
```

- The diffusivity $D(r)$ in eq. (9.1) is piecewise constant, with

$$
\begin{array}{lll}
D(r) = 0.13 & \begin{array}{l} 0 \le r \le (16-1)(0.25) \\ 0 \le r \le 3.75 \end{array} & \text{gray tissue} \\
D(r) = 0.65 & \begin{array}{l} (17-1)(0.25) \le r \le (86-1)(0.25) \\ 4 \le r \le 21.25 \end{array} & \text{white tissue} \\
D(r) = 0.13 & \begin{array}{l} (87-1)(0.25) \le r \le 25 \\ 21.50 \le r \le 25 \end{array} & \text{gray tissue}
\end{array}
$$

The boundary distances are half those used in [3] to reflect the use of spherical rather than Cartesian coordinates.

```
#
# Diffusivity
   D=rep(0,nr);
   for(i in 1:nr){
     if(i<=16){D[i]=0.13;}
     if((i>16)&(i<=86)){D[i]=0.65;}
     if(i> 86){D[i]=0.13;}
   }
```

The subscripts for D are hardcoded to provide a clear indication of the values of r at which $D(r)$ changes values. If the total number of grid points, nr=101, is changed, these coded values will also have to be changed accordingly.

- The numerical details of $D(r)$ are displayed if requested (ip = 2).

```
#
# Display D(r)
  if(ip==2){
    for(i in 1:nr){
      cat(sprintf("\n i = %3d  r = %4.2f  D(r) = %5.3f",
                  i,r[i],D[i]));
    }
  }
```

- The derivative $dD(r)/dr$ in eq. (9.1) is computed and placed in the vector dDdr. At the points of discontinuous change, i=16,86 ($r = 3.75, 21.25$), a two-point central finite difference (FD) approximation is used in place of the derivative (which is undefined at these two points)

$$\frac{\partial D}{\partial r}\Bigg|_i \approx \frac{D_{i+1} - D_{i-1}}{2\Delta r}$$

```
#
# dD/dr
  dDdr=rep(0,nr);
  for(i in 1:nr){
    if(i< 16){dDdr[i]=0;}
    if(i==16){dDdr[i]=(D[i+1]-D[i-1])/0.5;}
    if((i>16)&(i<86)){dDdr[i]=0;}
    if(i==86){dDdr[i]=(D[i+1]-D[i-1])/0.5;}
    if(i> 86){dDdr[i]=0;}
  }
```

The denominator of the FD approximation is $0.5 = (2)(0.25)$, that is, twice the spacing of the grid in r ($\Delta r = 0.25$).

- The numerical details of dD/dr are displayed if requested (ip = 2).

```
#
# Display dD/dr
  if(ip==2){
    for(i in 1:nr){
      cat(sprintf("\n i = %3d  r = %4.2f  dD/dr = %5.3f",
                  i,r[i],dDdr[i]));
    }
  }
```

The vectors D,dDdr are now available for use in the MOL/ODE routine discussed next. They do not require any special designation to be shared with the routine (a feature of R).

- IC (9.2a) is defined numerically, with $g(r) = \dfrac{1}{\sqrt{2\pi\epsilon}} e^{-0.5(r-12.5)^2/\epsilon}$ [3]. $g(r)$ is a Gaussian distribution centered at $r = 12.5$, the midpoint of the interval in r, $0 \leq r \leq 25$.

```
#
# Initial conditions
  u0=rep(0,nr);
  fact=1/((2*pi)^0.5*eps);
  for(i in 1:nr){
      u0[i]=exp(-0.5*(r[i]-12.5)^2/eps);
  }
  u0=fact*u0;
  ncall=0;
```

The statement u0=fact*u0 demonstrates the vector facility of R. In this case, u0 is a vector multiplied by a scalar, fact. Finally, the counter for the calls to the MOL/ODE routines is initialized.

- Selected output is displayed at the beginning of the numerical solution.

```
#
# Write selected parameters
  cat(sprintf("\n\n ncase = %2d",ncase));
#
# Write heading
  if(ip==1){
    cat(sprintf("\n Graphical output only\n"));
  }
```

- A vector of nout=6 output values of t is defined for $0 \leq t \leq 15$ so that tout has the values $t = 0, 3,...,15$ days.

```
#
# Independent variable for ODE integration
  nout=6;t0=0;tf=15;
  tout=seq(from=t0,to=tf,by=(tf-t0)/(nout-1));
```

- The nr=101 ODEs are programmed in pde_1 (discussed next) and integrated by lsodes (from deSolve). The ODE routine pde_1, the vector of output values tout and the IC vector u0 are the input to lsodes as expected (to define the MOL/ODE system). func,times,y are reserved names for lsodes.

```
#
```

```
# ODE integration
out=lsodes(y=u0,times=tout,func=pde_1)
nrow(out)
ncol(out)
```

The dimensions of the solution matrix out, out[6,101+1], are checked with the utilities nrow, ncol. The second dimension 102 includes a place for the ODE independent variable t as well as the 101 ODE dependent variables. The length of the IC vector u0 informs lsodes of the number of ODEs to be integrated (101).

- The numerical solution $u(r, t)$ of eq. (9.1) is placed in two arrays, u_plot1, u_plot2 for subsequent plotting. u_plot1 includes the solution for $t = 0$ while u_plot2 does not (the reason for this is explained subsequently).

```
#
# Arrays for plotting numerical solution
  u_plot1=matrix(0,nrow=nr,ncol=nout);
  u_plot2=matrix(0,nrow=nr,ncol=nout-1);
  for(it in 1:nout){
    for(i in 1:nr){
      u_plot1[i,it]=out[it,i+1];
      if(it>1){u_plot2[i,it-1]=u_plot1[i,it];}
    }
  }
```

The solution values for the interval in t are included in the matrices by a for with index it, and for the interval in r by a for with index i. The offset of 1 in out[it,i+1] is required since out[it,1] has the output values of t.

- The numerical solution in u_plot1 is displayed as a function of r and t for ip = 2.

```
#
# Display numerical solution
  if(ip==2){
    for(it in 1:nout){
      cat(sprintf(
      "\n    t        r      u(r,t)\n"));
      for(i in 1:nr){
        cat(sprintf("%5.1f%8.2f%10.3f\n",
        tout[it],r[i],u_plot1[i,it]));
      }
    }
  }
```

- The number of calls to the MOL/ODE routine pde_1 is displayed as a measure of the computational effort required to compute the numerical solution.

```
#
# Calls to ODE routine
  cat(sprintf("\n\n ncall = %5d\n\n",ncall));
```

- The numerical solution of eq. (9.1) is plotted with $t = 0$ included so that $t = 0, 3, 6,...,15$ and $t = 0$ (u_plot1) and not included so that $t = 3, 6,...,15$ (u_plot2). The resulting plots in Figs. 9.1a, 9.1b indicate that the Gaussian IC is a sharp pulse that quickly diffuses to much smaller values as emphasized in Fig. 9.1b by not including the IC. In other words, the Gaussian pulse simulates a point input (at $r = 12.5$ mm) of cancer cells that then disperse by diffusion and increase by the linear growth law of eqs. (9.3) (with ncase=1 set in the main program of Listing 9.1).

```
#
# Calls to ODE routine
  cat(sprintf("\n\n ncall = %5d\n\n",ncall));
#
# Plot u with t = 0
  par(mfrow=c(1,1));
  matplot(x=r,y=u_plot1,type="l",xlab="r",
          ylab="u(r,t),  t=0,3,...,15",xlim=c(0,r0),lty=1,
          main="u(r,t);  t=0,3,...,15;",lwd=2,col="black");
#
# Plot u without t = 0
  par(mfrow=c(1,1));
  matplot(x=r,y=u_plot2,type="l",xlab="r",
          ylab="u(r,t),  t=3,...,15",xlim=c(0,r0),lty=1,
          main="u(r,t);  t=3,...,15;",lwd=2,col="black");
```

Fig. 9.1a indicates the sharp Gaussian pulse at $t = 0$.

Fig. 9.1b indicates the dispersion of cells for $t > 0$, including the effect of the discontinuous change in $D(r)$, particularly near $r = 3.75$ as discussed previously. Specifically, the solution $u(r, t)$ has a larger (steeper) slope for $0 \leq r \leq 3.75$ where the diffusivity $D(r)$ is lower ($D(r) = 0.13$) in order to main continuity of the flux of cells. For $3.75 < r \leq 21.25$, the diffusivity is larger ($D(r) = 0.65$) so that the slope of $u(r, t)$ is lower. The numerical output from the main program of Listing 9.1 is discussed subsequently.

The units for the axes of Figs. 9.1a, 9.1b are determined by the units of $D(r)$ and ρ in eq. (9.1).

- $D(r)$: mm^2/day
- ρ: 1/day

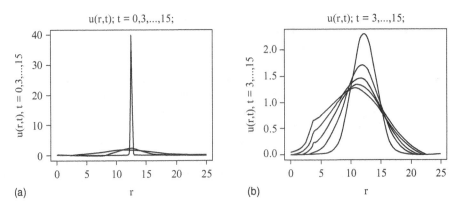

Figure 9.1 (a) $u(r,t)$ vs r with $t = 0, 3,...,15$, **ncase=1**, linear, (b) $u(r,t)$ vs r with $t = 3,...,15$, **ncase=1**, linear

In other words, r and t in the numerical solution of eq. (9.1), and in Figs. 9.1a, 9.1b, have the units mm and days, respectively.

This concludes the discussion of the main program. The MOL/ODE routine pde_1 called by lsodes is considered next.

(9.2.2) MOL/ODE routine

The programming of eqs. (9.1) to (9.3) is in Listing 9.2.

```
  pde_1=function(t,u,parms){
#
# Function pde_1 computes the t derivative vector of the
# u vector
#
# ur
  ur=dss004(0,r0,nr,u);
#
# Boundary conditions
  ur[1]=0;ur[nr]=0;
#
#
  nl=2;nu=2;
  urr=dss044(0,r0,nr,u,ur,nl,nu);
#
# PDE terms
  term1=rep(0,nr);term2=rep(0,nr);
  for(i in 1:nr){
    if(i==1){term1[i]=3*D[i]*urr[i]+(dDdr[i]*ur[i]);}
    if(i>1 ){term1[i]=D[i]*urr[i]+(dDdr[i]*ur[i])+
                      D[i]*(2/r[i])*ur[i];}
    if(ncase==1){term2[i]= rho*u[i];}
```

```
      if(ncase==2){term2[i]= rho*u[i]*(1-u[i]/umax);}
      if(ncase==3){term2[i]=-rho*u[i]*log2(exp(u[i]/exp(k/d)));}
  }
#
# PDE
  ut=rep(0,nr);
  for(i in 1:nr){
    ut[i]=term1[i]+term2[i];
  }
#
# Increment calls to pde_1
  ncall <<- ncall+1;
#
# Return derivative vector
  return(list(c(ut)));
}
```

Listing 9.2: MOL/ODE routine pde_1 for eqs. (9.1) to (9.3)

We can note the following details about Listing 9.2.

- The function is defined. The input arguments are: (1) the current value of t, (2) the vector of 101 ODE dependent variables, and (3) any parameters to be passed to pde_1 (unused).

  ```
    pde_1=function(t,u,parms){
  #
  # Function pde_1 computes the t derivative vector of the
  # u vector
  ```

- The derivative $\partial u/\partial r$ in eq. (9.1), ur, is computed by the library routine dss004 (accessed by a source in the main program in Listing 9.1). The output, ur, does not have to be declared (dimensioned) since this is done in dss004.

  ```
  #
  # ur
    ur=dss004(0,r0,nr,u);
  ```

- Eqs. (9.2b,c) are programmed as homogeneous Neumann BCs. Note the subscripts 1,nr corresponding to $r = 0, r_0$.

  ```
  #
  # Boundary conditions
    ur[1]=0;ur[nr]=0;
  ```

- The second derivative $\partial^2 u/\partial r^2$ in eq. (9.1), urr, is computed by dss044 (accessed by a source in the main program in Listing 9.1). nl=2,nu=2 specify Neumann BCs (nl=1,nu=1 would specify Dirichlet BCs for which the dependent variable $u(r,t)$ at the boundaries $r = 0, r_0$ is defined).

```
#
#
  nl=2;nu=2;
  urr=dss044(0,r0,nr,u,ur,nl,nu);
```

- The RHS terms of eq. (9.1) are computed.

```
#
# PDE terms
  term1=rep(0,nr);term2=rep(0,nr);
  for(i in 1:nr){
    if(i==1){term1[i]=3*D[i]*urr[i]+(dDdr[i]*ur[i]);}
    if(i>1 ){term1[i]=D[i]*urr[i]+(dDdr[i]*ur[i])+
                      D[i]*(2/r[i])*ur[i];}
    if(ncase==1){term2[i]= rho*u[i];}
    if(ncase==2){term2[i]= rho*u[i]*(1-u[i]/umax);}
    if(ncase==3){term2[i]=-rho*u[i]*log2(exp(u[i]/exp(k/d)));}
  }
```

The equivalence of the mathematical terms and the programming is explained next.

Terms in eq. (9.1)	Programming in pde_1
$D(r)\dfrac{\partial^2 u}{\partial r^2} + \dfrac{dD(r)}{dr}\dfrac{\partial u}{\partial r}$	`term1[i]=D[i]*urr[i]+(dDdr[i]*ur[i])`
$+D(r)\dfrac{2}{r}\dfrac{\partial u}{\partial r}, r \neq 0$	`+D[i]*(2/r[i])*ur[i];`
$3D(r)\dfrac{\partial^2 u}{\partial r^2} + \dfrac{dD(r)}{dr}\dfrac{\partial u}{\partial r}, r = 0$	`term1[i]=3D[i]*urr[i]+(dDdr[i]*ur[i])`
$f(u) = \rho u$ (ncase=1)	`term2[i]= rho*u[i];`
$\rho u(r,t)\left(1 - \dfrac{u(r,t)}{u_{max}}\right)$ (ncase=2)	`term2[i]= rho*u[i]*(1-u[i]/umax);`
$f(u) = -\rho u(r,t)\ln\left(\dfrac{u(r,t)}{e^{k/d}}\right)$ (ncase=3)	`term2[i]= -rho*u[i]`
	`*log2(exp(u[i]/exp(k/d)));`

We can note the following details about this comparison of the mathematics and coding for the RHS of eq. (9.1).

- The radial group $\dfrac{2}{r}\dfrac{\partial u}{\partial r}$ in eq. (9.1) is indeterminate at $r = 0$ with the form $0/0$ (the zero in the numerator follows from BC (9.1b)). This indeterminate form can be regularized with a straightforward application of l'Hospital's rule ([4], p359). The final result is

$$\lim_{r \to 0} \frac{2}{r} \frac{\partial u}{\partial r} = 2 \frac{\partial^2 u}{\partial r^2}$$

Thus, for $r = 0$ (i=1), the indeterminate form combines with the usual second derivative to give $3 \frac{\partial^2 u}{\partial r^2}$. Otherwise, for $r \neq 0$ (i>1), the radial group in eq. (9.1) is programmed directly (by using the vector r[i]).

- $f(u)$ in eq. (9.1) is programmed for the three cases ncase=1,2,3 according to eqs. (9.3).

- These two groups of terms are programmed as term1[i], term2[i], respectively, then added according to the RHS of eq. (9.1) to give the LHS derivative $\frac{\partial u}{\partial t}$ ($=$ ut).

This programming demonstrates the straightforward application of the MOL to PDEs with varying coefficients ($D(r)$) and changing source terms ($f(u)$). $D(r)$ can also be a function of u and t, and $f(u)$ can also be a function of r and t.

- Eq. (9.1) is programmed as discussed above to give the derivative vector ut.

```
#
# PDE
  ut=rep(0,nr);
  for(i in 1:nr){
    ut[i]=term1[i]+term2[i];
  }
```

- The number of calls to pde_1 is incremented, and the value is returned to the main program of Listing 9.1 by the <<- operator.

```
#
# Increment calls to pde_1
  ncall <<- ncall+1;
```

- The derivative vector ut is returned to lsodes by the combination of (1) c, the vector operator in R, (2) list (ode requires a list) and (3) return.

```
#
# Return derivative vector
  return(list(c(ut)));
}
```

The final } concludes pde_1.

The output from the routines of Listings 9.1, 9.2 follows.

(9.3) Model output

The numerical and graphical output for ncase=1,2,3 is reviewed next.

(9.3.1) Output for `ncase=1`**, linear**

The graphical output for `ncase=1` in the main program is in Figs. 9.1, b discussed previously. We consider here the numerical output (Table 9.1).

```
i =    1   r = 0.00   D(r) = 0.130
i =    2   r = 0.25   D(r) = 0.130
i =    3   r = 0.50   D(r) = 0.130
            .          .
            .          .
            .          .
   Output for r = 0.75 to 3.25
             removed
            .          .
            .          .
            .          .
i =   15   r = 3.50   D(r) = 0.130
i =   16   r = 3.75   D(r) = 0.130
i =   17   r = 4.00   D(r) = 0.650
i =   18   r = 4.25   D(r) = 0.650
            .          .
            .          .
            .          .
   Output for r = 4.50 to 20.75
             removed
            .          .
            .          .
            .          .
i =   85   r = 21.00  D(r) = 0.650
i =   86   r = 21.25  D(r) = 0.650
i =   87   r = 21.50  D(r) = 0.130
i =   88   r = 21.75  D(r) = 0.130
            .          .
            .          .
            .          .
   Output for r = 22.00 to 24.25
             removed
            .          .
            .          .
            .          .
```

Table 9.1: Abbreviated numerical output for `ncase=1`, linear

```
i =  99   r = 24.50   D(r) = 0.130
i = 100   r = 24.75   D(r) = 0.130
i = 101   r = 25.00   D(r) = 0.130


i =   1   r = 0.00   dD/dr = 0.000
i =   2   r = 0.25   dD/dr = 0.000
i =   3   r = 0.50   dD/dr = 0.000
              .          .
              .          .
              .          .

   Output for r = 0.75 to 3.00
             removed

              .          .
              .          .
              .          .

i =  14   r = 3.25   dD/dr = 0.000
i =  15   r = 3.50   dD/dr = 0.000
i =  16   r = 3.75   dD/dr = 1.040
i =  17   r = 4.00   dD/dr = 0.000
i =  18   r = 4.25   dD/dr = 0.000
              .          .
              .          .
              .          .

   Output for r = 4.50 to 20.50
             removed

              .          .
              .          .
              .          .

i =  84   r = 20.75   dD/dr = 0.000
i =  85   r = 21.00   dD/dr = 0.000
i =  86   r = 21.25   dD/dr = -1.040
i =  87   r = 21.50   dD/dr = 0.000
i =  88   r = 21.75   dD/dr = 0.000
              .          .
              .          .
              .          .

   Output for r = 22.00 to 24.25
             removed

              .          .
              .          .
              .          .
```

Table 9.1: (*Continued*)

```
i =  99   r = 24.50   dD/dr = 0.000
i = 100   r = 24.75   dD/dr = 0.000
i = 101   r = 25.00   dD/dr = 0.000

ncase =  1

[1] 6

[1] 102
      t         r      u(r,t)
     0.0     0.00      0.000
     0.0     0.25      0.000
     0.0     0.50      0.000
               .         .
               .         .
               .         .
  Output for r = 0.75 to
        11.50 removed
               .         .
               .         .
               .         .
     0.0    11.75      0.000
     0.0    12.00      0.000
     0.0    12.25      1.753
     0.0    12.50     39.894
     0.0    12.75      1.753
     0.0    13.00      0.000
     0.0    13.25      0.000
               .         .
               .         .
               .         .
  Output for r = 13.50 to
        24.25 removed
               .         .
               .         .
               .         .
     0.0    24.50      0.000
     0.0    24.75      0.000
     0.0    25.00      0.000
               .         .
               .         .
               .         .
```

Table 9.1: (*Continued*)

```
Output for t = 3 to 12
         removed

            .              .
            .              .
            .              .
   t         r        u(r,t)
 15.0      0.00       0.038
 15.0      0.25       0.039
 15.0      0.50       0.043

            .              .
            .              .
            .              .
Output for r = 0.75 to
       10.75 removed

            .              .
            .              .
            .              .
 15.0     11.00        1.259
 15.0     11.25        1.253
 15.0     11.50        1.243
 15.0     11.75        1.231
 15.0     12.00        1.215
 15.0     12.25        1.195
 15.0     12.50        1.173
 15.0     12.75        1.149
 15.0     13.00        1.121

            .              .
            .              .
            .              .
Output for r = 13.25 to
       23.75 removed

            .              .
            .              .
            .              .
 15.0     24.00        0.002
 15.0     24.25        0.001
 15.0     24.50        0.001
 15.0     24.75        0.001
 15.0     25.00        0.001

ncall =    302
```

Table 9.1: (*Continued*)

We can note the following details about this output.

- $D(r)$ has discontinuous changes at $r = 3.75, 21.25$ (i=16,86) as programmed in the main program of Listing 9.1.

```
i =   16   r = 3.75   D(r) = 0.130
i =   17   r = 4.00   D(r) = 0.650

i =   86   r = 21.25  D(r) = 0.650
i =   87   r = 21.50  D(r) = 0.130
```

- dD/dr has nonzero values at $r = 3.75, 21.25$ (i=16,86) as programmed in the main program of Listing 9.1.

```
i =   15   r = 3.50    dD/dr = 0.000
i =   16   r = 3.75    dD/dr = 1.040
i =   17   r = 4.00    dD/dr = 0.000

i =   85   r = 21.00   dD/dr = 0.000
i =   86   r = 21.25   dD/dr = -1.040
i =   87   r = 21.50   dD/dr = 0.000
```

These nonzero values of dD/dr result from the second-order, centered FD approximations programmed in Listing 9.1.

- The ODE solution matrix out has the dimensions out[6,102] as expected. Again, the second dimension is 101 + 1 = 102 to include the current value of t as well as the 101 ODE dependent variables.
ncase = 1
[1] 6
[1] 102

- The Gaussian IC $g(r) = \dfrac{1}{\sqrt{2\pi\epsilon}} e^{-0.5(r-12.5)^2/\epsilon}$ centered at $r = 12.50$ has the numerical values

```
0.0    12.00       0.000
0.0    12.25       1.753
0.0    12.50      39.894
0.0    12.75       1.753
0.0    13.00       0.000
```

This sharp pulse is displayed in Fig. 9.1a, and can be interpreted as an initial source or concentration of cancer cells at $r = 12.5$.

- The initial source is then dispersed by a combination of the RHS diffusion and linear growth terms (for ncase=1) in eq. (9.1). For example, at $t = 15$ days, the distribution near $r = 12.50$ is

```
15.0    11.00       1.259
15.0    11.25       1.253
```

15.0	11.50	1.243
15.0	11.75	1.231
15.0	12.00	1.215
15.0	12.25	1.195
15.0	12.50	1.173
15.0	12.75	1.149
15.0	13.00	1.121

Note the shift in the peak cell density toward lower values of r as reflected in Fig. 9.1b.

- The computational effort was modest with ncall = 302.

In general, the numerical output reflects the programming in Listings 9.1 and 9.2 for ncase=1, and the graphical output in Figs. 9.1a,b.

(9.3.2) Output for ncase=2, logistic

Abbreviated numerical output for ncase = 2 is given in Table 9.2.

For $t = 0$, the IC is again the Gaussian distribution. The solution for $t = 15$ is similar to the solution for ncase=1. This close agreement is explained by considering the logistic function (from eqs. (9.3) for ncase=2).

$$f(u) = \rho u(r,t) \left(1 - \frac{u(r,t)}{u_{max}}\right)$$

t	r	u(r,t)
0.0	11.75	0.000
0.0	12.00	0.000
0.0	12.25	1.753
0.0	12.50	39.894
0.0	12.75	1.753
0.0	13.00	0.000
0.0	13.25	0.000

t	r	u(r,t)
15.0	11.00	1.253
15.0	11.25	1.247
15.0	11.50	1.237
15.0	11.75	1.225
15.0	12.00	1.209
15.0	12.25	1.190
15.0	12.50	1.168
15.0	12.75	1.143
15.0	13.00	1.116

Table 9.2: Abbreviated numerical output for ncase=2, logistic

This function has two important limiting cases.

- $u(r, t) << u_{max}$ so that $f(u) \approx \rho u(r, t)$, that is, approximately the linear function of ncase=1.

- $u(r, t) \approx u_{max}$ so that $f(u) \approx 0$, and the logistic growth function is close to zero. In other words, u_{max} is a limiting value for $u(r, t)$ (although $u(r, t) > u_{max}$ is possible so that $f(u) < 0$ in which case logistic growth becomes logistic depletion).

In the present case, $u_{max} = 62.5$ from [3], so the first condition applies ($u(r, t = 15) < 2$ from Table 9.2) and the logistic function for ncase=2 is close to the linear function for ncase=1, and therefore the two solutions are also close. For this reason, the graphical output is similar to Figs. 9.1a,b and is not included here to conserve space. To test this analysis of the logistic function, the reader can execute the routines in Listings 9.1, 9.2 with a smaller value of u_{max}.

(9.3.3) Output for ncase=3, Gompertz

Abbreviated numerical output for ncase = 3 is given in Table 9.3.

For $t = 0$, the IC is again the Gaussian distribution. The solution at $t = 15$ is substantially different than for ncase=2 (compare Tables 9.2 and 9.3). This is

```
ncase =  3

[1] 6

[1] 102

     t          r       u(r,t)
   0.0       0.00       0.000
   0.0       0.25       0.000
   0.0       0.50       0.000
               .           .
               .           .
               .           .

Output for r = 0.75 to
        11.50 removed

               .           .
               .           .
               .           .

   0.0      11.75       0.000
   0.0      12.00       0.000
   0.0      12.25       1.753
   0.0      12.50      39.894
   0.0      12.75       1.753
   0.0      13.00       0.000
   0.0      13.25       0.000
```

Table 9.3: Abbreviated numerical output for ncase=3, Gompertz

```
          .           .
          .           .
          .           .
Output for r = 13.50 to
      24.25 removed
          .           .
          .           .
          .           .
0.0    24.50        0.000
0.0    24.75        0.000
0.0    25.00        0.000
          .           .
          .           .
          .           .
Output for t = 3 to 12
         removed
          .           .
          .           .
          .           .
  t        r        u(r,t)
15.0    0.00        0.030
15.0    0.25        0.031
15.0    0.50        0.034
          .           .
          .           .
          .           .
Output for r = 0.75 to
      10.75 removed
          .           .
          .           .
          .           .
15.0   11.00        0.915
15.0   11.25        0.910
15.0   11.50        0.904
15.0   11.75        0.895
15.0   12.00        0.884
15.0   12.25        0.870
15.0   12.50        0.855
15.0   12.75        0.837
15.0   13.00        0.818
          .           .
          .           .
          .           .
```

Table 9.3: (*Continued*)

```
Output for r = 13.25 to
        23.75 removed

            .              .
            .              .
            .              .
15.0     24.00         0.002
15.0     24.25         0.001
15.0     24.50         0.001
15.0     24.75         0.001
15.0     25.00         0.000

ncall =    302
```

Table 9.3: (*Continued*)

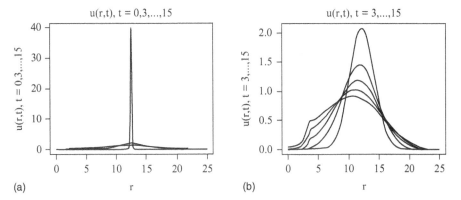

u(r,t), t = 0,3,...,15   u(r,t), t = 3,...,15

(a)   r   (b)   r

Figure 9.2 (a) $u(r,t)$ vs r with $t = 0, 3,...,15$, **ncase=3**, Gompertz, (b) $u(r,t)$ vs r with $t = 3,...,15$, **ncase=3**, Gompertz

confirmed by comparing Figs. 9.1b and 9.2b (note the difference in the ordinate (vertical) scales). Thus, with the particular sets of parameters used in Listing 9.1, the Gompertz function offers an alternative to the logistic function (eqs. (9.3)) for comparison with observed growth rates and cell densities. Again, the computational effort for ncase=3 is modest with ncall = 302.

As a final case, we can consider changing the statement for $D(r)$ in Listing 9.1 from

```
if((i>16)&(i<=86)){D[i]=0.65;}
```

to

```
if((i>16)&(i<=86)){D[i]=0.13;}
```

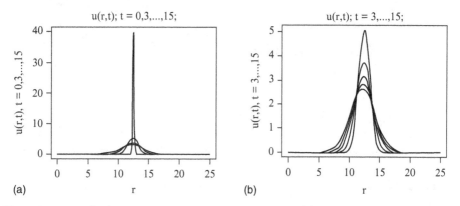

Figure 9.3 (a) $u(r,t)$ vs r with $t = 0, 3,...,15$, ncase=1, $D(r) = 0.13$, (b) $u(r,t)$ vs r with $t = 3,...,15$, ncase=1, $D(r) = 0.13$

so that $D(r)$ is constant in r at 0.13. This should be reflected in a symmetric response to the Gaussian IC, and this conclusion is confirmed in Figs. 9.3a,b.

(9.4) *p*-Refinement error analysis

The numerical solutions discussed previously cannot be evaluated for accuracy with an analytical solution, except possibly for the linear case ncase=1, $D(r) = 0.13$, for which an analytical solution would be available, but it would be complicated. Thus an evaluation is required that does not include the use of an analytical solution. This can be done in two ways:

1. The number of grid points in r could be changed (from nr=101) and the effect on the numerical solution observed, which is usually termed h-refinement since the grid interval in the numerical analysis literature is often denoted with h. Changing the number of grid points would require reprogramming for the arrays D,dDdr in the main program of Listing 9.1 to retain the change in $D(r)$ at $r = 3.75, 21.25$.
2. The order of the FD approximations of the derivatives in r in eq. (9.1) can be changed, which is usually termed p-refinement since the order of the approximations of derivatives in the numerical analysis literature is often denoted with p.

Here we briefly consider p-refinement, by using dss006, dss046 (with $p = 6$) in pde_1 of Listing 9.2 in place of dss004, dss044 (with $p = 4$), that is,

```
ur=dss004(0,r0,nr,u);
```

changed to

```
ur=dss006(0,r0,nr,u);
```

and

```
urr=dss044(0,r0,nr,u,ur,nl,nu);
```

changed to

```
urr=dss046(0,r0,nr,u,ur,nl,nu);
```

The numerical output for the fourth and sixth order differentiators is compared next.

```
dss004,dss044

15.0    11.00       1.259
15.0    11.25       1.253
15.0    11.50       1.243
15.0    11.75       1.231
15.0    12.00       1.215
15.0    12.25       1.195
15.0    12.50       1.173
15.0    12.75       1.149
15.0    13.00       1.121

dss006,dss046

15.0    11.00       1.259
15.0    11.25       1.253
15.0    11.50       1.243
15.0    11.75       1.231
15.0    12.00       1.215
15.0    12.25       1.195
15.0    12.50       1.173
15.0    12.75       1.149
15.0    13.00       1.121
```

The numerical output is the same, implying that the truncation error[2] for the two orders, $c_4(\Delta r)^4$ and $c_6(\Delta r)^6$, is negligibly small. This also implies Δr is small enough that the truncation errors are negligible so increasing the number of grid points in r, that is, reducing Δr (h-refinement), will not improve the accuracy of the solutions.

   With this brief error analysis, we can conclude that the accuracy of the numerical solutions is apparently better than four significant figures. However, this is not a proof of this accuracy, but only that there appears to be spatial convergence (with respect to r) of the solutions to four figures (the solutions could still have large errors, for example, from a programming error in pde_1).

---

[2]FD approximations are based on a truncated Taylor series. The resulting truncation error of the approximations is of the form $c(\Delta r)^p$ where c is a constant, Δr is the grid spacing in r ($\Delta r = 0.25$ in the preceding discussion) and p is the order of the approximation.

(9.5) Summary and conclusions

The model of eqs. (9.1) to (9.3) for cell density dynamics of glioblastomas demonstrates: (1) the numerical solution of PDEs with variable coefficients ($D(r)$ as a function of r to quantify different diffusion rates in gray and white tissue), (2) three established growth rates: linear, logistic and Gompertz and (3) the use of spherical coordinates, including the regularization of a singularity ($1/r$ for $r \rightarrow 0$). The spatial convergence of the solution was also considered through p-refinement of the FD approximations of the derivatives in r. The MOL implementation of the model is straightforward and computationally efficient.

References

[1] American Brain Tumor Association. http://www.abta.org/brain-tumor-information/types-of-tumors/glioblastoma.html.

[2] Jackson, P.R., et al. (2015), Patient-specific mathematical neuro-oncology: using a simple proliferation and invasion tumor model to inform clinical practice, *Bull. Math. Biol.*, **77**, 5, 846–856.

[3] Ozugurlu, E. (2015), A note on the numerical approach for the reaction-diffusion problem to model the density of the tumor growth dynamics, *Comput. Math. Appl.*, **69**, 1504–1517

[4] Schiesser, W.E., and G.W. Griffiths (2009), *A Compendium of Partial Differential Equation Models*, Cambridge University Press, Cambridge, UK

10

MOL ANALYSIS WITH A VARIABLE GRID: ANTIGEN-ANTIBODY BINDING KINETICS

This chapter pertains to the method of lines (MOL) solution of an ODE/PDE model on a variable grid. The ODE/PDE application is used to illustrate the possible advantage of a variable grid, that is, the concentration of spatial grid points in regions where the solution varies rapidly to improve the spatial resolution of the solution.

Specifically, the intent of the chapter is to:

- Present an ODE/PDE model for antigen-antibody binding kinetics including the required initial conditions (ICs) and boundary conditions (BCs).
- Present the algorithms and associated routines for a variable spatial grid MOL solution of the ODE/PDE model.
- Compare the numerical solutions computed with uniform and variable spatial grids.
- Discuss the possible advantage of a variable spatial grid, and the increased complexity of a variable grid implementation.

The ODE/PDE model is presented next, followed by a MOL solution programmed in R.

(10.1) ODE/PDE model

We now consider the ODE/PDE model, taken originally from [5], pertaining to the transport and binding kinetics of an analyte, e.g., an antigen, on an antibody surface of a fiber-optic biosensor.

Method of Lines PDE Analysis in Biomedical Science and Engineering, First Edition. William E. Schiesser.
© 2016 John Wiley & Sons, Inc. Published 2016 by John Wiley & Sons, Inc.
Companion website: www.wiley.com/go/Schiesser/PDE_Analysis

The PDE defining the antigen concentration $c(z,t)$[1] is the classical 1D diffusion equation (Fick's second law for mass diffusion) in Cartesian coordinates.

$$\frac{\partial c}{\partial t} = D \frac{\partial^2 c}{\partial z^2} \tag{10.1}$$

Eq. (10.1) is second order in z and therefore requires two boundary conditions (BCs), which are taken as

$$D\frac{\partial c(z=0,t)}{\partial z} = k_f c(z=0,t)(c_{b,sat} - c_b) - k_r c_b \tag{10.2a}$$

$$c(z=h,t) = c_{bulk} \tag{10.2b}$$

Eq. (10.2a) describes the binding of an antigen at the antibody surface corresponding to $z = 0$, as depicted in Figure 10.1. The LHS of eq. (10.2a), $D\dfrac{\partial c(z=0,t)}{\partial z}$, is the rate of diffusion of the antigen to the antibody surface at $z = 0$ according to Fick's first law for mass diffusion. The RHS of eq. (10.2a), $k_f c(z=0,t)(c_{b,sat} - c_b) - k_r c_b$, is the difference between the forward rate of adsorption (or binding) of the antigen, $k_f c(z = 0,t)(c_{b,sat} - c_b)$, and the rate of desorption (or unbinding), $k_r c_b$. c_b is the concentration of the antigen bound to the antibody and is defined by an ODE (considered next); k_f and k_r are mass transfer constants (or rate constants) for the forward and reverse binding, respectively. Note that the forward rate goes to zero as the bound concentration, c_b, reaches a saturation value, $c_{b,sat}$; the rate of reverse binding is proportional to the bound concentration, c_b.

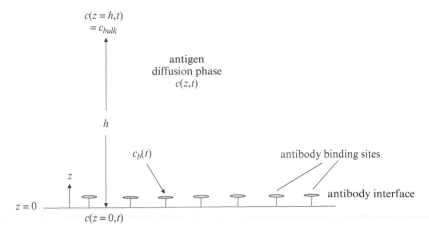

Figure 10.1   Schematic of diffusion-binding system

[1] In the numerical analysis literature, an ODE dependent variable is frequently denoted with y and a PDE dependent variable is denoted with u. Rather than attempt to use this different terminology for the ODE/PDE model, we use c for both the ODE and PDE.

Eq. (10.1) is first order in t and therefore requires one IC.

$$c(z, t = 0) = 0 \tag{10.3}$$

Eq. (10.2b) indicates that at a sufficiently large distance from the antibody interface, $z = h$, the antigen concentration has the constant value c_{bulk}. In fact, this value of c_{bulk} drives the model away from the homogeneous (zero) IC of eq. (10.3).

$c_b(t)$, the concentration of the antigen bound on the antibody, is given by the ODE

$$\frac{dc_b}{dt} = k_f c(z = 0, t)(c_{b,sat} - c_b) - k_r c_b \tag{10.4}$$

A homogeneous IC is also specified for eq. (10.4).

$$c_b(t = 0) = 0 \tag{10.5}$$

Eqs. (10.1) to (10.5) constitute the complete ODE/PDE model. The model variables and parameters are summarized in Table 10.1. These variables have SI (MKS) units. Also, the concentration of the bound component at the antibody interface is expressed through an area concentration (e.g., c_b is in mols/m^2) while the diffusing antigen is expressed through a volume (or bulk) concentration (e.g., c is in mols/m^3). The parameter numerical values and units are listed in Table 10.2 [5,3].

We now consider the coding of eqs. (10.1) to (10.5) with emphasis on the use of a nonuniform grid in z.

Variable	Description
c	antigen fluid concentration, mols/m^3
z	distance from antibody interface, m
t	time, s
c_b	concentration of the bound antigen, mols/m^2

Table 10.1: Dependent and independent variables of eqs. (10.1) and (10.4).

Parameter	Units, numerical value
D	1×10^{-10}m^2/s
k_f	1×10^5M^{-1}s^{-1}(M = molarity = mols/m^3)
k_r	1×10^{-2}1/s
$c_{b,sat}$	2.66×10^{-8}mols/m^2
c_{bulk}	4.48×10^{-5}mols/m^3
h	5.0×10^{-5}m
c_0	0 mols/m^3
c_{b0}	0 mols/m^2

Table 10.2: Parameters and numerical values for eqs. (10.1) to (10.5)

(10.2) MOL routines

The main program for eqs. (10.1) to (10.5) follows.

(10.2.1) Main program

```
#
#   Variable grid MOL
#
#   Antigen-antibody binding
#
# Delete previous workspaces
  rm(list=ls(all=TRUE))
#
# Access ODE integrator
  library("deSolve");
#
# Access files
  setwd("g:/chap10");
  source("pde_1.R") ;
  source("dss032.R");source("dss32a.R");
#
# Level of output
#
#   Detailed output - ip = 1
#
#   Graphical output only - ip = 2
#
  ip=1;
#
# Type of grid in z
#
#   Uniform  - iuv = 1
#
#   Variable - iuv = 2
#
  iuv=1;
#
# Parameter numerical values
  D=1.0e-10; kf=1.0e+05; kr=1.0e+01;
  cbulk=4.48e-05; cbsat=1.66e-09;
  h=5.0e-05; c0=0; cb0=0;
#
# Spatial grid
  zl=0; zu=5.0e-05; nz=21; dz=(zu-zl)/(nz-1);
#
```

```
# z grid
  z=rep(0,nz);
#
# Uniform grid
  if(iuv==1){
    z=seq(from=zl,to=zu,by=(zu-zl)/(nz-1));
  }
#
# Variable grid
  if(iuv==2){
    for(i in 1:nz){
      cat(sprintf("\n i = %2d   z[i] = %10.5e",
                  i,z[i]));
      if(i<21){z[i+1]=z[i]+0.0047619*i*zu};
    }
  }
#
# Initial conditions
  u0=rep(0,nz+1);
  for(i in 1:nz){
    u0[i]=c0;
  }
  u0[nz+1]=cb0;
#
# Independent variable for ODE integration
  t0=0;tf=100;nout=51;
  tout=seq(from=t0,to=tf,by=(tf-t0)/(nout-1));
  ncall=1;
#
# uz, uzz approximations
  igrid=rep(0,nz);
  igrid[1]=1;    igrid[2]=2;
  igrid[nz-1]=4;igrid[nz]=5;
  for(i in 3:(nz-2)){
    igrid[i]=3;
  }
#
# ODE integration
  out=ode(func=pde_1,times=tout,y=u0);
#
# Store numerical solution
  c=matrix(0,nrow=nout,ncol=nz);
  cb=rep(0,nout);
  for(it in 1:nout){
    for(i in 1:nz){
      c[it,i]=out[it,i+1];
```

```
    }
    cb[it]=out[it,nz+2];
  }
#
# Calls to ODE routine
  cat(sprintf("\n ncall = %4d\n",ncall));
#
# Display grid
  if(ip==1){
    cat(sprintf("\n      i        z"));
    for(i in 1:nz){
      cat(sprintf("\n %5.0d %12.3e",i,z[i]));
    }
  }
#
# Display numerical solution at z = 0
  if(ip==1){
    cat(sprintf(
      "\n      t     c(z=0,t)        cb(t)"));
    for(it in 1:nout){
      cat(sprintf("\n %6.0f %12.4e %12.4e",
                  tout[it],c[it,1],cb[it]));
    }
  }
#
# Plot z grid
  par(mfrow=c(1,1))
  iplot=rep(0,nz);
  iplot=seq(from=1,to=nz,by=1);
  plot(iplot,z,xlab="i",ylab="z",
     main="z vs i",pch="o",lwd=2);
  lines(iplot,z,type="l",lwd=2);
#
# Plot z grid spacing
  par(mfrow=c(1,1))
  zs=rep(0,(nz-1));
  for(i in 1:(nz-1)){
    zs[i]=z[i+1]-z[i];
  }
   plot(iplot[1:(nz-1)],zs,xlab="i",ylab="zs",
      main="z spacing vs i",pch="o",lwd=2);
  lines(iplot[1:(nz-1)],zs,type="l",lwd=2);
#
# Plot c(z=0,t)
  par(mfrow=c(1,1))
  plot(tout,c[,1],xlab="t",ylab="c(z=0,t)",
```

```
   main="c(z=0,t)",type="l",lwd=2);
# points(tout,c[,1],pch="o",lwd=2);
#
# Plot cb(t)
  par(mfrow=c(1,1))
  plot(tout,cb,xlab="t",ylab="cb(t)",
    main="cb(t)",type="l",lwd=2);
# points(tout,cb,pch="o",lwd=2);
```

Listing 10.1: Main program for eqs. (10.1) to (10.5)

We can note the following details about Listing 10.1.

- Previous workspaces are removed

```
#
#     Variable grid MOL
#
#     Antigen-antibody binding
#
# Delete previous workspaces
  rm(list=ls(all=TRUE))
```

- The ODE library, with ode used later, is accessed

```
#
# Access ODE integrator
  library("deSolve");
#
# Access files
  setwd("g:/chap10");
  source("pde_1.R") ;
  source("dss032.R");source("dss32a.R");
```

The setwd (set working directory) requires editing for the local computer. Note that / is used in place of the usual \. In addition to the MOL/ODE routine, pde_1 (discussed subsequently), the differentiation routines dss032, dss32a for a variable spatial grid are accessed.

- The level of output is selected. For ip=1 the solution is displayed numerically and graphically.

```
#
# Level of output
#
#     Detailed output - ip = 1
#
```

```
#    Graphical output only - ip = 2
#
  ip=1;
```

- The type of spatial grid is selected. We start with a uniform grid (iuv=1), then subsequently proceed to a variable grid.

```
#
# Type of grid in z
#
#    Uniform  - iuv = 1
#
#    Variable - iuv = 2
#
  iuv=1;
```

- The model parameters are defined numerically.

```
#
# Parameter numerical values
  D=1.0e-10; kf=1.0e+05; kr=1.0e+01;
  cbulk=4.48e-05; cbsat=1.66e-09;
  h=5.0e-05; c0=0; cb0=0;
```

- The grid in z is defined on 21 points for the interval $z_l \le z \le z_u$ with spacing dz.

```
#
# Spatial grid
  zl=0; zu=5.0e-05; nz=21; dz=(zu-zl)/(nz-1);
```

- The grid in z is defined numerically for the uniform case, iuv=1, and the variable case, uuv=2. For iuv=1, the utility seq defines a sequence of z values with spacing $(h - 0)/(nz - 1) = (5.0 \times 10^{-5} - 0)/(21 - 1) = 0.25 \times 10^{-5}$.

```
#
# z grid
  z=rep(0,nz);
#
# Uniform grid
  if(iuv==1){
    z=seq(from=zl,to=zu,by=(zu-zl)/(nz-1));
  }
#
# Variable grid
  if(iuv==2){
    for(i in 1:nz){
```

```
         cat(sprintf("\n i = %2d   z[i] = %10.5e",
                     i,z[i]));
         if(i<21){z[i+1]=z[i]+0.0047619*i*zu{;
      }
   }
```

For the variable case, iuv=2, the grid points are spaced as a linear function of the index i. The constant 0.0047619 was determined by trial-and-error so that z[nz] = z[21] = zu = 5.0e-05[2].

- The ICs for the PDE (eqs. (10.1) and (10.3)) and the ODE (eqs. (10.4) and (10.5)) are placed in the vector u0 (of length nz+1 = 22).

```
#
# Initial conditions
  u0=rep(0,nz+1);
  for(i in 1:nz){
    u0[i]=c0;
  }
  u0[nz+1]=cb0;
```

- The grid of output values of t is defined for the interval $0 \le t \le 100$ with 51 points (including $t = 0$), so the output values are $t = 0, 2, \ldots, 100$.

```
#
# Independent variable for ODE integration
  t0=0;tf=100;nout=51;
  tout=seq(from=t0,to=tf,by=(tf-t0)/(nout-1));
  ncall=1;
```

ncall=1 is used for the initial calculation of the weighting coefficients of the spatial derivative approximations[3].

- The type of approximation of the spatial derivatives at each grid point in z is defined.

```
#
# uz, uzz approximations
  igrid=rep(0,nz);
  igrid[1]=1;    igrid[2]=2;
  igrid[nz-1]=4;igrid[nz]=5;
  for(i in 3:(nz-2)){
    igrid[i]=3;
  }
```

---

[2]The trial-and-error converged rapidly to a five-figure value. Other algorithms for definition of the variable grid could be used at this point.
[3]The spatial derivative approximations for the variable grid are discussed in the chapter appendix.

Briefly,

- – `igrid[1]=1`: At `i=1`, point `i=1` and four points to the right, `i=2,3,4,5`, are used in the derivative approximations.
- – `igrid[2]=2`: At `i=2`, one point to the left, `i=1`, and three points to the right, `i=3,4,5`, are used in the derivative approximations.
- – `igrid[nz-1]=4`: At `i=nz-1`, one point to the right, `i=nz`, and three points to left, `i=nz-2,nz-3,nz-4`, are used in the derivative approximations.
- – `igrid[nz]=5`: At `i=nz`, point `i=nz` and four points to the left, `i=nz-1,nz-2,nz-3,nz-4`, are used in the derivative approximations.
- – `igrid[i]=3`: At the intermediate points `i=3,4,...,nz-2`, two points to the left, `i-1,i-2`, and two points to the right, `i+1,i+2`, are used in the derivative approximations.[4]

- The `nz + 1 = 21 + 1 = 22` ODEs are integrated by ode. As expected, the RHS inputs are (1) the MOL/ODE routine pde_1 (discussed next), (2) the IC vector u0, and (3) the vector of output values of t, tout. func, times, y are reserved names. The number of ODEs to be integrated is defined by the length of the vector u0.

```
#
# ODE integration
  out=ode(func=pde_1,times=tout,y=u0);
```

- The solution matrix out from ode is placed in a matrix, c, and a vector, cb with a for in t (index it) and a for in z (index i). The offset of 1 in out[it,i+1], out[it,nz+2] is required since out[it,1] is reserved for the values of t (in tout).

```
#
# Store numerical solution
  c=matrix(0,nrow=nout,ncol=nz);
  cb=rep(0,nout);
  for(it in 1:nout){
    for(i in 1:nz){
       c[it,i]=out[it,i+1];
    }
    cb[it]=out[it,nz+2];
  }
```

- The total number of calls to pde_1 is displayed at the end of the solution as a measure of the computational effort required to compute the solution.

```
#
# Calls to ODE routine
  cat(sprintf("\n ncall = %4d\n",ncall));
```

---

[4]The spatial derivative approximations for igrid=1,2,3,4,5 are derived and discussed in the chapter appendix.

- For ip=1, the points in the spatial grid z are displayed.

```
#
# Display grid
  if(ip==1){
    cat(sprintf("\n       i        z"));
    for(i in 1:nz){
      cat(sprintf("\n %5.0d %12.3e",i,z[i]));
    }
  }
```

The uniform and variable grids (iuv=1,2) are discussed subsequently.

- The interface concentrations $c(z = 0, t), c_b(t)$ are displayed as a function of t (with the for in it). These concentrations are of particular interest in elucidating the transfer of the antigen at the interface $z = 0$, which is the reason for concentrating the grid points near $z = 0$ for the variable grid case.

```
#
# Display numerical solution at z = 0
  if(ip==1){
    cat(sprintf(
      "\n       t     c(z=0,t)        cb(t)"));
    for(it in 1:nout){
      cat(sprintf("\n %6.0f %12.4e %12.4e",
                  tout[it],c[it,1],cb[it]));
    }
  }
```

- The grid in z is displayed graphically. The index i, $1 \le i \le 21$, in vector iplot is defined by a seq.

```
#
# Plot z grid
  par(mfrow=c(1,1))
  iplot=rep(0,nz);
  iplot=seq(from=1,to=nz,by=1);
  plot(iplot,z,xlab="i",ylab="z",
     main="z vs i",pch="o",lwd=2);
  lines(iplot,z,type="l",lwd=2);
```

- The spacing in the z grid, vector zs, is displayed graphically to demonstrate the difference between the uniform and variable grids.

```
#
# Plot z grid spacing
  par(mfrow=c(1,1))
  zs=rep(0,(nz-1));
```

```
for(i in 1:(nz-1)){
  zs[i]=z[i+1]-z[i];
}
 plot(iplot[1:(nz-1)],zs,xlab="i",ylab="zs",
   main="z spacing vs i",pch="o",lwd=2);
lines(iplot[1:(nz-1)],zs,type="l",lwd=2);
```

zs has the difference between successive grid points, zs[i]=z[i+1]-z[i].

- The solution of eq. (10.1), $c(z = 0, t)$, is plotted against t. This concentration is in c[,1] where the values for $t = 0, 2, \ldots, 100$ have been included with the (,) subscript.

```
#
# Plot c(z=0,t)
  par(mfrow=c(1,1))
  plot(tout,c[,1],xlab="t",ylab="c(z=0,t)",
    main="c(z=0,t)",type="l",lwd=2);
# points(tout,c[,1],pch="o",lwd=2);
```

- The solution of eq. (10.4), $c_b(t)$, is plotted against t.

```
#
# Plot cb(t)
  par(mfrow=c(1,1))
  plot(tout,cb,xlab="t",ylab="cb(t)",
    main="cb(t)",type="l",lwd=2);
# points(tout,cb,pch="o",lwd=2);
```

This completes the discussion of the main program in Listing 10.1. The numerical and graphical output for the uniform and variable grids is discussed later. The MOL/ODE routine pde_1 called by ode is considered next.

(10.2.2) MOL/ODE routine

The MOL routine for eqs. (10.1), (10.2), (10.4) follows.

```
  pde_1=function(t,u,parms){
#
# Function pde_1 computes the t derivative vector
# of the u vector
#
# ODE and PDE
  c=rep(0,nz);
  for(i in 1:nz){
    c[i]=u[i];
  }
```

```
  cb=u[nz+1];
#
# BC
  c[nz]=cbulk;
#
# cz
  cz=dss032(ncall,npts,igrid,nz,z,c);
#
# BC
  cz[1]=(1/D)*(kf*c[1]*(cbsat-cb)-kr*cb);
#
# czz
  czz=dss032(ncall,npts,igrid,nz,z,cz);
#
# PDE
  ct=D*czz;
  ct[nz]=0;
#
# ODE
  cbt=kf*c[1]*(cbsat-cb)-kr*cb;
#
# Derivative vector
  ut=rep(0,nz+1);
  for(i in 1:nz){
    ut[i]=ct[i];
  }
  ut[nz+1]=cbt;
#
# Increment calls to pde_1
  ncall<<-ncall+1;
#
# Return derivative vector
  return(list(c(ut)));
}
```

Listing 10.2: MOL/ODE routine for eqs. (10.1), (10.2), (10.4)

We can note the following details about Listing 10.2.

- The function is defined. u is the vector of 22 ODE dependent variables at a particular time t. parms for passing parameters to pde_1 is unused.

```
  pde_1=function(t,u,parms){
#
# Function pde_1 computes the t derivative vector
# of the u vector
```

- u is placed in c,cb to facilitate the programming of eqs. (10.1) and (10.4).

```
#
# ODE and PDE
  c=rep(0,nz);
  for(i in 1:nz){
    c[i]=u[i];
  }
  cb=u[nz+1];
```

- Dirichlet BC (10.2b) is programmed (for point i=nz).

```
#
# BC
  c[nz]=cbulk;
```

- The derivative $\partial c(z,t)/\partial z$ in eq. (10.1) is computed by a call to dss032.

```
#
# cz
  cz=dss032(ncall,npts,igrid,nz,z,c);
```

The input and output arguments of dss032 are:

- ncall: For ncall=1, the weighting coefficients in the spatial derivative approximations in dss032 are computed. For ncall>1 (from the statement at the end of dss032 that increments ncall), the weighting coefficients for ncall=1 are used repeatedly (these details are discussed in the chapter appendix).
- npts: The number of points used in the derivative spatial approximations (npts=5).
- igrid: Definition of the type of spatial derivative approximations (as explained previously after Listing 10.1).
- nz: Total number of grid points in z, e.g., nz=21.
- z: Vector of grid points defined in the main program of Listing 10.1 (for a uniform or variable grid, iuv=1,2).
- c: The vector to be differentiated numerically.
- cz: The derivative of c with respect to z. cz does not have to be dimensioned since this is done in dss032.
- BC (10.2a) is programmed for i=1 to define the derivative $\partial c(z=0,t)/\partial z$.

```
#
# BC
  cz[1]=(1/D)*(kf*c[1]*(cbsat-cb)-kr*cb);
```

BC (10.2a) is of the third type (or a Robin BC) since $\partial c(z=0,t)/\partial z$ is a function of $c(z=0,t)$ (c[1]). Since $c_b(t)$ is used in eq. (10.2a), the link between eqs. (10.1) and (10.4) is established (the product c[1]*cb also demonstrates a nonlinear BC).

- The second derivative in eq. (10.1), $\partial^2 c(z,t)/\partial z^2$, is computed by differentiating the first derivative, $\partial c(z,t)/\partial z$.

```
#
# czz
  czz=dss032(ncall,npts,igrid,nz,z,cz);
```

- Eq. (10.1) is programmed using the vector utility of R (czz and ct are 21-vectors, but subscripting is not required). In particular, D*czz is a scalar-vector product (each element of czz is multiplied by D).

```
#
# PDE
  ct=D*czz;
  ct[nz]=0;
```

BC (10.2b) defines $c(z = h, t)$, so the derivative in t is set to zero (to prevent the ODE integrator ode from moving $c(z = h, t)$ away from its prescribed value in BC (10.2b)).

- Eq. (10.4) is programmed.

```
#
# ODE
  cbt=kf*c[1]*(cbsat-cb)-kr*cb;
```

- $\partial c/\partial t$ and dc_b/dt are placed in a single vector ut for return to ODE integrator ode.

```
#
# Derivative vector
  ut=rep(0,nz+1);
  for(i in 1:nz){
    ut[i]=ct[i];
  }
  ut[nz+1]=cbt;
```

- The number of calls to pde_1 is incremented and returned to the main program of Listing 10.1 via the <<- operator.

```
#
# Increment calls to pde_1
  ncall<<-ncall+1;
```

This change in ncall instructs dss032 to calculate the weighting coefficients in the spatial derivative approximations only one time (as explained in the chapter appendix).

- The derivative vector ut is returned to ode through a combination of c, the R vector operator, list to return a list as required by ode, and return.

```
#
# Return derivative vector
  return(list(c(ut)));
}
```

The final } concludes ode_1.

This completes the programming of eqs. (10.1) to (10.5). Numerical and graphical (plotted) output is considered next.

(10.3) Model output

Abbreviated numerical output (from ip=1 in Listing 10.1), and graphical output for the uniform grid iuv=1 are considered first.

(10.3.1) Uniform grid

The numerical output includes the grid z and the solutions to eqs. (10.1) and (10.4), $c(z = 0, t), c_b(t)$.

We can note the following details about this output.

- The interval in z is $z_l(= 0) \leq z \leq z_u(= 5.0 \times 10^{-5})$ with a spacing of 0.25×10^{-5}.
- The ICs of eqs. (10.3) and (10.5) are confirmed.

```
   t     c(z=0,t)        cb(t)
   0   0.0000e+00    0.0000e+00
```

This check is worthwhile since if the ICs are incorrect, the numerical solution will be incorrect.

- The interval in t is $0 \leq t \leq 100$ with an output increment of 2 for 51 output points (including $t = 0$).
- The solution is in five-figure agreement with a previously reported solution, [3], Chapter 2.
- The computational effort is modest with ncall = 310.

The graphical output is in Figs. 10.2a to 10.2d.

Fig. 10.2a confirms the linear increase in the grid z,

Fig. 10.2b confirms the uniform spacing 0.25×10^{-5} in the grid z.

Fig. 10.2c,d indicate that $c(z = 0, t), c_b(t)$ start off at homogeneous (zero) ICs and approach a steady state (equilibrium) as reflected in Table 10.3. Also, the solutions are positive and monotonic (free of oscillation) as expected for the physical problem.

The output for the case of a uniform grid provides a basis for a comparative analysis of the output for the case of a variable grid, as explained next.

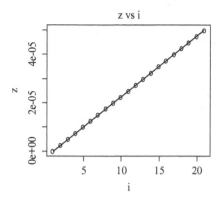

Figure 10.2a Uniform spatial grid

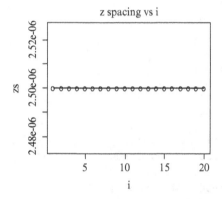

Figure 10.2b Increment of uniform spatial grid

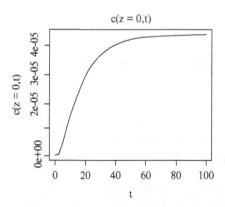

Figure 10.2c $c(z = 0, t)$ vs t, uniform grid

```
ncall =   310

       i        z
       1     0.000e+00
       2     2.500e-06
       3     5.000e-06
       4     7.500e-06
       5     1.000e-05
       6     1.250e-05
       7     1.500e-05
       8     1.750e-05
       9     2.000e-05
      10     2.250e-05
      11     2.500e-05
      12     2.750e-05
      13     3.000e-05
      14     3.250e-05
      15     3.500e-05
      16     3.750e-05
      17     4.000e-05
      18     4.250e-05
      19     4.500e-05
      20     4.750e-05
      21     5.000e-05

       t     c(z=0,t)         cb(t)
       0     0.0000e+00     0.0000e+00
       2     3.5319e-07     4.8262e-12
       4     3.0023e-06     4.5580e-11
       6     6.9498e-06     1.0483e-10
       8     1.1089e-05     1.6306e-10
      10     1.5090e-05     2.1540e-10
                .                .
                .                .
                .                .
Output for t =12 to 88 removed
                .                .
                .                .
                .                .
      90     4.4692e-05     5.1273e-10
      92     4.4706e-05     5.1284e-10
      94     4.4718e-05     5.1294e-10
      96     4.4729e-05     5.1302e-10
      98     4.4738e-05     5.1310e-10
     100     4.4746e-05     5.1316e-10
```

Table 10.3: Abbreviated numerical output for the uniform grid

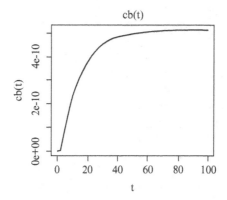

Figure 10.2d $c_b(t)$ vs t, uniform grid

(10.3.2) Variable grid

The variable grid, which concentrates the grid points in the neighborhood of the inter-face at $z = 0$, is implemented by changing iuv=1 to iuv=2 in the main program of Listing 10.1. The resulting output is summarized next.

We can note the following details about this output.

```
ncall =   343
      i          z
      1      0.000e+00
      2      2.381e-07
      3      7.143e-07
      4      1.429e-06
      5      2.381e-06
      6      3.571e-06
      7      5.000e-06
      8      6.667e-06
      9      8.571e-06
     10      1.071e-05
     11      1.310e-05
     12      1.571e-05
     13      1.857e-05
     14      2.167e-05
     15      2.500e-05
     16      2.857e-05
     17      3.238e-05
     18      3.643e-05
     19      4.071e-05
```

Table 10.4: Abbreviated numerical output for the variable grid

```
20      4.524e-05
21      5.000e-05

 t      c(z=0,t)         cb(t)
 0      0.0000e+00     0.0000e+00
 2      3.6104e-07     4.9911e-12
 4      2.9423e-06     4.4670e-11
 6      6.9045e-06     1.0412e-10
 8      1.1084e-05     1.6302e-10
10      1.5107e-05     2.1554e-10
            .              .
            .              .
            .              .

Output for t =12 to 88 removed

            .              .
            .              .
            .              .

 90     4.4702e-05     5.1281e-10
 92     4.4714e-05     5.1291e-10
 94     4.4725e-05     5.1300e-10
 96     4.4735e-05     5.1307e-10
 98     4.4743e-05     5.1314e-10
100     4.4751e-05     5.1320e-10
```

Table 10.4: (*Continued*)

- The interval in z is $z_l(= 0) \leq z \leq z_u(= 5.0 \times 10^{-5})$ with a variable spacing ranging from $2.381e-07 - 0.000e+00 = 2.381 \times 10^{-7}$ at $z = 0$ to $5.000e-05 - 4.524e-05 = 4.76 \times 10^{-6}$ at $z = 5.0 \times 10^{-5}$, a variation in the spacing by more than a factor of 10. Also, the point 21 5.000e-05 confirms the use of the scaling factor 0.0047619 in if(i<21)z[i+1]=z[i]+0.0047619*i*zu (from Listing 10.1).

- The ICs of eqs. (10.3) and (10.5) are confirmed.

```
 t      c(z=0,t)         cb(t)
 0      0.0000e+00     0.0000e+00
```

- The solution is in at least three-figure agreement with the previous solution (for iuv=1) for $96 \leq t \leq 100$. This is also true for $0 \leq t \leq 4$ if the small values of $c(z = 0, t), c_b(t)$ are considered.

Uniform grid

```
 t      c(z=0,t)         cb(t)
 0      0.0000e+00     0.0000e+00
 2      3.5319e-07     4.8262e-12
 4      3.0023e-06     4.5580e-11
```

```
        .               .
        .               .
        .               .
  96  4.4729e-05    5.1302e-10
  98  4.4738e-05    5.1310e-10
 100  4.4746e-05    5.1316e-10

Variable grid
    t      c(z=0,t)        cb(t)
    0   0.0000e+00    0.0000e+00
    2   3.6104e-07    4.9911e-12
    4   2.9423e-06    4.4670e-11
        .               .
        .               .
        .               .
   96  4.4735e-05    5.1307e-10
   98  4.4743e-05    5.1314e-10
  100  4.4751e-05    5.1320e-10
```

- The computational effort is modest with ncall = 343, so the variable grid does not require a substantial increase in the calls to pde_1. Also, within pde_1, the required calculations are approximately the same since the weighting coefficients for the approximate spatial derivatives are calculated only once (for ncall=1), then used repeatedly in subsequent calls to pde_1.

In summary, the variable grid for the particular application to eqs. (10.1) to (10.5) gives comparable results to the uniform grid with a finer spatial resolution near $z = 0$.

The graphical output is in Figs. 10.3a to 10.3d.

Fig. 10.3a confirms the nonlinear increase in z, with small changes in z concentrated near $z = 0$.

Fig. 10.3b confirms the variable spacing in z, which in this case is linear in i due to the function 0.0047619*i*zu (from Listing 10.1)).

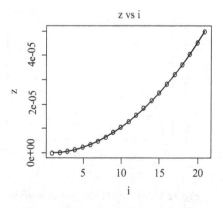

Figure 10.3a   Variable spatial grid

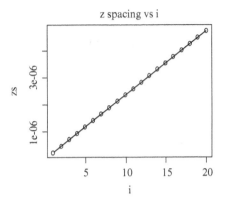

Figure 10.3b   Incremental spacing of variable spatial grid

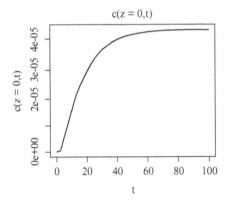

Figure 10.3c $c(z = 0, t)$ vs t, variable grid

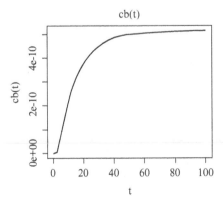

Figure 10.3d $c_b(t)$ vs t, variable grid

Fig. 10.3c is indistinguishable from Fig. 10.2c since $c(z = 0, t)$ for the uniform and variable grids is essentially the same.

Fig. 10.3d is indistinguishable from Fig. 10.2d since $c_b(t)$ for the uniform and variable grids is essentially the same.

(10.4) Summary and conclusions

The variable grid implemented in dss032/dss32a is general purpose in the sense that the analyst can locate the grid points to concentrate them where they are required (to enhance spatial resolution). However, this use of the variable grid requires some attention.

- The grid points can be located by inspection, but they must be single-valued with continuously increasing or decreasing values. A function to meet these requirements (e.g., if(i<21)z[i+1]=z[i]+0.0047619*i*zu) generally facilitates the placement of the points.
- The variation of the grid spacing should be smooth. That is, large, discontinuous changes in the spacing, or a zero spacing (interval), should be avoided. Again, a smooth function can be used to ensure this condition.
- The total interval in z should be confirmed (as was done with the factor 0.0047619).
- The total number of grid points (e.g., nz=21) is not restricted, but the total computational effort to produce a MOL solution will increase with the number of points, so this value should be selected judiciously. Also, comparative execution of the routines for different numbers of grid points is advised to ensure the number is adequate (for a spatially converged solution of acceptable accuracy) without an excessively large value for the total grid points.
- Extension of the use of dss032/dss32a to 2D and 3D applications with multiple PDEs is straightforward, as illustrated for uniform grids [4].

Variations on the use of dss032/dss32a are possible. For example,

- If the specification of the approximations is changed to (from Listing 10.1)
```
#
# uz approximations
  igrid=rep(0,nz);
  igrid[1]=1;  igrid[2]=2;
  igrid[3]=3; igrid[nz]=5;
  for(i in 4:(nz-1)){
    igrid[i]=4;
  }
```
then at the interior points i=4,5,...i,...,nz-1, the approximation is based on three points to the left of i and one point to the right. This approximation is termed a five-point biased upwind (5pbu) finite difference (FD) and is generally useful for RHS convective PDE terms of the form $-v\dfrac{\partial u}{\partial z}$ for $v > 0$ (flow left to right in z).

- If the specification of the approximations is changed to (from Listing 10.1)

```
#
# uz approximations
  igrid=rep(0,nz);
  igrid[1]=1;   igrid[nz-2]=3;
  igrid[nz-1]=4; igrid[nz]=5;
  for(i in 2:(nz-3)){
    igrid[i]=2;
  }
```

  then at the interior points i=2,5,...i,...,nz-3, the approximation is based
  on one point to the left of i and three points to the right. This approximation is
  also termed a 5pbu FD approximation and is generally useful for RHS convective
  PDE terms of the form $-v\dfrac{\partial u}{\partial z}$ for $v < 0$ (flow right to left in z). The 5pbu FD on a
  uniform grid is implemented in function dss020 for $v > 0$ and $v < 0$, as discussed
  in Chapter 1.

- If a 5pbu approximation is used, it requires prior knowledge of the sign of the
  velocity (the direction of flow). Also, although this approximation is generally
  effective for strongly convective (hyperbolic) PDEs, it will not accommodate discontinuities and sharp spatial changes in the PDE solution (in this case, a special
  nonlinear approximation is required such as a flux limiter). This point of selecting
  an approximation for first-order convective derivatives is discussed in Chapter 1.

- For second-order RHS parabolic (diffusion) PDE terms of the form $D\dfrac{\partial^2 u}{\partial z^2}$ the
  centered approximation corresponding igrid[i]=3 should be used at the inte-
  rior points (as illustrated in pde_1 of Listing 10.2). In other words, some form
  of upwinding is generally required for the approximation of first-order hyperbolic
  (convective) derivatives while centered approximations are used for second-order
  parabolic (diffusive) derivatives.

- With the solutions to the model PDEs available, additional functions can be computed and displayed that provide insight into the features of the numerical solution.
  For example, a key variable of the model is the antigen transfer rate at $z = 0$, which
  can be computed (from eq. (10.2a)) as $k_f c(z = 0, t)(c_{b,sat} - c_b) - k_r c_b$ from the
  available solutions $c(z = 0, t), c_b$. Additionally, the individual terms in this rate can
  be calculated and displayed, e.g., $k_f c(z = 0, t)(c_{b,sat} - c_b)$ for the forward (fluid
  to antibody) binding rate, as depicted in Fig. 10.1, to demonstrate the nonlinear
  logistic rate, and $-k_r c_b$ for the reverse unbinding rate. Since these terms involve a
  subset of parameters for the full model, this analysis can identify the contributions
  of individual parameters that can then be judged quantitatively.

- A variable grid is not limited to the Lagrange interpolation polynomial as discussed
  in the chapter appendix. For example, splines can be used, which have important
  continuity properties at the grid points, and routines are available in R for splines,
  but this approach will not be discussed here because of limited space.

Generally the use of a variable grid can be developed using the concepts and examples discussed in the chapter appendix. The approximations for spatial derivatives (1) of any order, (2) with any number of grid points (3), defined on uniform or variable grids, (4) based on Lagrange interpolation polynomials, can be defined numerically (i.e., by the computed weighting coefficients) with readily available routines [2]. These approximations can be applied to systems of linear and nonlinear PDEs, in 1D, 2D and 3D, with a variety of BC types. Thus, MOL analysis on variable grids to achieve improved spatial resolution is an open-ended procedure applicable to a broad spectrum of PDE models and applications.

APPENDIX: VARIABLE GRID ANALYSIS

The numerical differentiators in dss032 used in the ODE/PDE routine of Listing 10.2 are derived and tested in this appendix.

(A.10.1) Derivation of numerical differentiators

The starting point for the derivation of the numerical differentiators is a set of independent/dependent variable pairs, $(z, f(z))$. z represents the spatial variable z in eq. (10.1), and $f(z)$ represents the dependent variable $u(z, t)$ of eq. (10.1) (for a particular value of t). n pairs are represented in a table.

The first requirement is to develop a function approximating $f(z)$ by using the $(z, f(z))$ pairs in Table A.10.1. The approximating function can then be differentiated to give the numerical differentiator.

The approximating function for $n = 5$ is taken as a fourth-order Lagrange interpolation polynomial based on the points z_1, z_2, z_3, z_4, z_5.

$$
\begin{aligned}
f(z) \approx \; & \frac{(z - z_2)(z - z_3)(z - z_4)(z - z_5)}{(z_1 - z_2)(z_1 - z_3)(z_1 - z_4)(z_1 - z_5)} f(z_1) \\[6pt]
& + \frac{(z - z_1)(z - z_3)(z - z_4)(z - z_5)}{(z_2 - z_1)(z_2 - z_3)(z_2 - z_4)(z_2 - z_5)} f(z_2) \\[6pt]
& + \frac{(z - z_1)(z - z_2)(z - z_4)(z - z_5)}{(z_3 - z_1)(z_3 - z_2)(z_3 - z_4)(z_3 - z_5)} f(z_3) \qquad \text{(A.10.1)} \\[6pt]
& + \frac{(z - z_1)(z - z_2)(z - z_3)(z - z_5)}{(z_4 - z_1)(z_4 - z_2)(z_4 - z_3)(z_4 - z_5)} f(z_4) \\[6pt]
& + \frac{(z - z_1)(z - z_2)(z - z_3)(z - z_4)}{(z_5 - z_1)(z_5 - z_2)(z_5 - z_3)(z_5 - z_4)} f(z_5)
\end{aligned}
$$

Note that the numerator of each term in eq. (A.10.1) is a fourth-order polynomial in z, e.g., in the first $(f(z_1))$ term, $(z - z_2)(z - z_3)(z - z_4)(z - z_5)$. $f(z)$ of eq. (A.10.1) is

$$
\begin{array}{cc}
z_1 & f(z_1) \\
z_2 & f(z_2) \\
\cdot & \cdot \\
\cdot & \cdot \\
\cdot & \cdot \\
z_n & f(z_n)
\end{array}
$$

Table A.10.1: Prescribed $(z, f(z))$ pairs

then approximated at any particular value of z by using the entries from Table A.10.1 (for $n = 5$).

$f(z)$ of eq. (A.10.1) can then be differentiated with respect to z to give the first derivative $df(z)/dz$

$$
\frac{df(z)}{dz} = \frac{\begin{aligned}(z - z_2)(z - z_3)(z - z_4) + (z - z_2)(z - z_3)(z - z_5) + \\ (z - z_2)(z - z_4)(z - z_5) + (z - z_3)(z - z_4)(z - z_5)\end{aligned}}{(z_1 - z_2)(z_1 - z_3)(z_1 - z_4)(z_1 - z_5)} f(z_1)
$$

$$
+ \frac{\begin{aligned}(z - z_1)(z - z_3)(z - z_4) + (z - z_1)(z - z_3)(z - z_5) + \\ (z - z_1)(z - z_4)(z - z_5) + (z - z_3)(z - z_4)(z - z_5)\end{aligned}}{(z_2 - z_1)(z_2 - z_3)(z_2 - z_4)(z_2 - z_5)} f(z_2)
$$

$$
+ \frac{\begin{aligned}(z - z_1)(z - z_2)(z - z_4) + (z - z_1)(z - z_2)(z - z_5) + \\ (z - z_1)(z - z_4)(z - z_5) + (z - z_2)(z - z_4)(z - z_5)\end{aligned}}{(z_3 - z_1)(z_3 - z_2)(z_3 - z_4)(z_3 - z_5)} f(z_3) \quad \text{(A.10.2)}
$$

$$
+ \frac{\begin{aligned}(z - z_1)(z - z_2)(z - z_3) + (z - z_1)(z - z_2)(z - z_5) + \\ (z - z_1)(z - z_3)(z - z_5) + (z - z_2)(z - z_3)(z - z_5)\end{aligned}}{(z_4 - z_1)(z_4 - z_2)(z_4 - z_3)(z_4 - z_5)} f(z_4)
$$

$$
+ \frac{\begin{aligned}(z - z_1)(z - z_2)(z - z_3) + (z - z_1)(z - z_2)(z - z_4) + \\ (z - z_1)(z - z_3)(z - z_4) + (z - z_2)(z - z_3)(z - z_4)\end{aligned}}{(z_5 - z_1)(z_5 - z_2)(z_5 - z_3)(z_5 - z_4)} f(z_5)
$$

Note that the numerator of each term of eq. (A.10.2) is a third-order polynomial in z, e.g., for the $f(z_1)$ term, $(z - z_2)(z - z_3)(z - z_4) + (z - z_2)(z - z_3)(z - z_5) + (z - z_2)(z - z_4)(z - z_5) + (z - z_3)(z - z_4)(z - z_5)$. Eq. (A.10.2) can be used to calculate $df(z)/dz$ for a particular value of z. The only requirement to compute the derivative is the use of the entries in Table A.10.1 for $n = 5$.

If $z = z_1$ in eq. (A.10.2), the formula for the first derivative at $z = z_1$, $df(z_1)/dz$, results.

$$
\frac{df(z_1)}{dz} = \frac{\begin{aligned}(z_1 - z_2)(z_1 - z_3)(z_1 - z_4) + (z_1 - z_2)(z_1 - z_3)(z_1 - z_5) + \\ (z_1 - z_2)(z_1 - z_4)(z_1 - z_5) + (z_1 - z_3)(z_1 - z_4)(z_1 - z_5)\end{aligned}}{(z_1 - z_2)(z_1 - z_3)(z_1 - z_4)(z_1 - z_5)} f(z_1)
$$

$$+ \frac{(z_1 - z_3)(z_1 - z_4)(z_1 - z_5)}{(z_2 - z_1)(z_2 - z_3)(z_2 - z_4)(z_2 - z_5)} f(z_2)$$

$$+ \frac{(z_1 - z_2)(z_1 - z_4)(z_1 - z_5)}{(z_3 - z_1)(z_3 - z_2)(z_3 - z_4)(z_3 - z_5)} f(z_3) \qquad \text{(A.10.3)}$$

$$+ \frac{(z_1 - z_2)(z_1 - z_3)(z_1 - z_5)}{(z_4 - z_1)(z_4 - z_2)(z_4 - z_3)(z_4 - z_5)} f(z_4)$$

$$+ \frac{(z_1 - z_2)(z_1 - z_3)(z_1 - z_4)}{(z_5 - z_1)(z_5 - z_2)(z_5 - z_3)(z_5 - z_4)} f(z_5)$$

Similarly, the substitutions $z = z_2$, $z = z_3$, $z = z_4$, and $z = z_5$ in eq. (A.10.2) gives respectively the differentiation formulas for $df(z_2)/dz$, $df(z_3)/dz$, $df(z_4)/dz$, and $df(z_5)/dz$.

Note that the factors which multiply $f(z_1)$, $f(z_2)$, $f(z_3)$, $f(z_4)$ and $f(z_5)$ in eq. (A.10.3) are just constants once the values of z_1, z_2, z_3, z_4 and z_5 are defined. Thus, eq. (A.10.3) can be written in the general form

$$df(z_1))/dz = c_1 f(z_1) + c_2 f(z_2) + c_3 f(z_3) + c_4 f(z_4) + c_5 f(z_5) \qquad \text{(A.10.4)}$$

where c_1, c_2, c_3, c_4 and c_5 are weighting coefficients (calculated in dss32a and passed to dss032 as the third argument, vector c).

If the following substitutions for a uniform grid with spacing dz are made in eq. (A.10.3)

$$dz = (z_2 - z_1) = (z_3 - z_2) = (z_4 - z_3) = (z_5 - z_4)$$

$$2dz = (z_3 - z_1) = (z_4 - z_2) = (z_5 - z_3)$$

$$3dz = (z_4 - z_1) = (z_5 - z_2)$$

$$4dz = (z_5 - z_1) \qquad \text{(A.10.5)}$$

$$-dz = (z_1 - z_2) = (z_2 - z_3) = (z_3 - z_4) = (z_4 - z_5)$$

$$-2dz = (z_1 - z_3) = (z_2 - z_4) = (z_3 - z_5)$$

$$-3dz = (z_1 - z_4) = (z_2 - z_5)$$

$$-4dz = (z_1 - z_5)$$

Eq. (A.10.3) reduces to the following five-point differentiation formula for $df(z_1)/dz$ for a uniform grid.

$$df(z_1)/dz = \frac{\begin{array}{c}((-1)(-2)(-3) + (-1)(-2)(-4) + \\ (-1)(-3)(-4) + (-2)(-3)(-4))\end{array}}{(-1)(-2)(-3)(-4)} f(z_1)/dz$$

$$+ \frac{(-2)(-3)(-4)}{(1)(-1)(-2)(-3)} f(z_2)/dz$$

$$+ \frac{(-1)(-3)(-4)}{(2)(1)(-1)(-2)} f(z_3)/dz$$

$$+ \frac{(-1)(-2)(-4)}{(3)(2)(1)(-1)} f(z_4)/dz$$

$$+ \frac{(-1)(-2)(-3)}{(4)(3)(2)(1)} f(z_5)/dz$$

or after numerically evaluating the coefficients of the $f(z_1)$ to $f(z_5)$ terms (with $(1)(2)(3)(4) = 4!$)

$$df(z_1)/dz = (1/4!dz)(-50f(z_1) + 96f(z_2) - 72f(z_3) + 32f(z_4) - 6f(z_5))$$

$$(A.10.6)$$

Equations for $df(z_2)/dz, df(z_3)/dz, df(z_4)/dz, df(z_5)/dz$ analogous to eq. (A.10.6) follow from substituting successively $z = z_2, z_3, z_4, z_5$ in eq. (A.10.2). The resulting system of five equations can be summarized as

$$
\begin{bmatrix} df(z_1)/dz \\ df(z_2)dz \\ df(z_3)/dz \\ df(z_4)/dz \\ df(z_5)/dz \end{bmatrix} = \frac{1}{4!dz} \begin{bmatrix} -50 & 96 & -72 & 32 & -6 \\ -6 & -20 & 36 & -12 & 2 \\ 2 & -16 & 0 & 16 & -2 \\ -2 & 12 & -36 & 20 & 6 \\ 6 & -32 & 72 & -96 & 50 \end{bmatrix} \begin{bmatrix} f(z_1) \\ f(z_2) \\ f(z_3) \\ f(z_4) \\ f(z_5) \end{bmatrix}
\qquad (A.10.7)
$$

The 5×5 coefficient matrix is a Bickley differentiation matrix for $n = 4, m = 1, p = 0, 1, 2, 3, 4$ [1]. The RHS is a matrix-vector multiplication with dimensions $(5 \times 5)(5 \times 1) = (5 \times 1)$. The final result is the vector $[df(z_1)/dz, \ldots , df(z_5)/dz]^T$, that is,

$$df(z_1)/dz = \frac{1}{4!dz}(-50f(z_1) + 96f(z_2) - 72f(z_3) + 32f(z_4) - 6f(z_5))$$

$$df(z_2)/dz = \frac{1}{4!dz}(-6f(z_1) - 20f(z_2) + 36f(z_3) - 12f(z_4) + 2f(z_5))$$

$$df(z_3)/dz = \frac{1}{4!dz}(2f(z_1) - 16f(z_2) + 0f(z_3) + 16f(z_4) - 2f(z_5))$$

$$df(z_4)/dz = \frac{1}{4!dz}(-2f(z_1) + 12f(z_2) - 36f(z_3) + 20f(z_4) + 6f(z_5))$$

$$df(z_5)/dz = \frac{1}{4!dz}(6f(z_1) - 32f(z_2) + 72f(z_3) - 96f(z_4) + 50f(z_5))$$

so that each derivative $df(z_i)/dz, i = 1, \ldots, 5$ is approximated as a weighted sum (linear combination) of the functional values $[f(z_1), \ldots, f(z_5)]^T$. The same is true for the variable grid case (iuv=2 in Listing 10.1). In both cases (uniform and variable), the weighting coefficients are calculated in dss32a only once with ncall=1.

In summary, the numerical derivatives for a grid of five points in z can be calculated from eq. (A.10.2) for a variable grid (the values of z are not necessarily uniformly spaced as in Figs. 10.3a,b), and from eq. (A.10.7) for a uniform grid (the values of z are uniformly spaced as in Figs. 10.2a,b).

(A.10.2) Testing of numerical differentiators

Two tests of eqs. (A.10.2) and (A.10.7) are now considered.

(A.10.2.1) Differentiation matrix

The 5×5 numerical differentiation matrix of eq. (A.10.7) can be confirmed by the following test program.

```
rm(list=ls(all=TRUE))
setwd("g:/chap10")
source("dss32a.R");
fact=1*2*3*4;
z=seq(from=1,to=5,by=1);
for(itype in 1:5){
  cat(sprintf("\n\n itype = %2d",itype));
  coeff=dss32a(itype,z);
  coeff=fact*coeff;
  cat(sprintf("\n%6.1f %6.1f %6.1f %6.1f %6.1f",
    coeff[1],coeff[2],coeff[3],coeff[4],coeff[5]));
}
```

Listing A.10.1: Test program for dss32a

We can note the following details about Listing A.10.1.

- Previous workspaces are cleared.

  ```
  rm(list=ls(all=TRUE))
  ```

- dss32a is accessed with the source utility. The setwd (set working directory) requires editing for the local computer. Note the use of / rather than the usual \.

  ```
  setwd("g:/chap10")
  source("dss32a.R");
  ```

- 4! is required subsequently.

  ```
  fact=1*2*3*4;
  ```

- A 5-vector is defined with the values $z = 1, 2, 3, 4, 5$ (so that the grid in z has a uniform spacing of 1.

```
z=seq(from=1,to=5,by=1);
```

- For each of five approximations from eq. (A.10.2) (for(itype in 1:5)), dss32a is called to implement eq. (A.10.2).

```
for(itype in 1:5){
  cat(sprintf("\n\n itype = %2d",itype));
  coeff=dss32a(itype,z);
  coeff=fact*coeff;
  cat(sprintf("\n%6.1f %6.1f %6.1f %6.1f %6.1f",
    coeff[1],coeff[2],coeff[3],coeff[4],coeff[5]));
}
```

The resulting 5-vector of weighting coefficients, coeff, (as illustrated in eq. (A.10.3) for itype=1) is displayed. Note the multiplication by fact so that the output can be compared with the 5×5 differentiation matrix of eq. (A.10.7). This multiplication demonstrates the R vector facility, that is, multiplication of a vector by a scalar (with no subscripting).

The output from the test program of Listing A.10.1 follows. The differentiation matrix of eq. (A.10.7) is confirmed.

```
itype =  1
-50.0    96.0  -72.0    32.0    -6.0

itype =  2
 -6.0  -20.0    36.0  -12.0     2.0

itype =  3
  2.0  -16.0     0.0    16.0    -2.0

itype =  4
 -2.0    12.0  -36.0    20.0     6.0

itype =  5
  6.0  -32.0    72.0  -96.0    50.0
```

Table A.10.2: Numerical output from the test program of Listing A.10.1

(A.10.2.2) Test functions

With the preceding confirmation of dss32a, dss032 can now be tested. Specifically, the numerical derivatives from eqs. (A.10.2) and (A.10.7) can be tested using functions with known analytical derivatives. This is done with the following program.

```
#
# Test problems for variable grid
#
#   nfcn=1: Polynomial of varying order
#
#   nfcn=2: Sine function
#
# Remove previous work spaces
  rm(list=ls(all=TRUE))
#
# Access dss032, dss32a
  setwd("g:/chap10");
  source("dss032.R");
  source("dss32a.R");
#
# Problem, grid parameters
  npts=5;n=21;nfcn=1;ncall=1;
#
# Arrays for test problem
  z=rep(0,n);igrid=rep(0,n);
  u=rep(0,n);uzze=rep(0,n);diff=rep(0,n);
#
# Display selected parameters
  cat(sprintf("\n nfcn = %2d   n = %3d\n\n",nfcn,n));
#
# Polynomial of varying order (selected with #)
  if(nfcn==1){
    p=1.5;
    for(i in 1:n){
      z[i]=(0.05*(i-1))^p;
#     u[i]=1;
#     u[i]=z[i];
#     u[i]=z[i]^2;
#     u[i]=z[i]^3;
      u[i]=z[i]^4;
#     u[i]=z[i]^5;
#     u[i]=z[i]^6;
    }
#
#   Definition of approximations
    igrid[1]=1;  igrid[2]=2;
    igrid[n-1]=4;igrid[n]=5;
    for(i in 3:(n-2)){
      igrid[i]=3;
    }
#
```

```
#     Numerical uz, uzz
      uz=dss032(ncall,npts,igrid,n,z,u);
      uzz=dss032(ncall,npts,igrid,n,z,uz);
#
#     Exact uzz (selected with #)
      cat(sprintf("\n      i  igrid      z          u
        uzz        uzze       diff"));
      for(i in 1:n){
#     uzze[i]=0;
#     uzze[i]=2;
#     uzze[i]=6*z[i];
      uzze[i]=12*z[i]^2;
#     uzze[i]=20*z[i]^3;
#     uzze[i]=30*z[i]^4;
      diff[i]=uzz[i]-uzze[i];
#
#     Display numerical and exact derivatives, difference
      cat(sprintf("\n %5d %5d %8.3f %8.3f %8.3f %8.3f %10.5f",
        i,igrid[i],z[i],u[i],uzz[i],uzze[i],diff[i]));
      }
#
# nfcn=1 complete
  }
#
# Sine function
  if(nfcn==2){
    p=1.5
    for(i in 1:n){
      z[i]=(0.05*(i-1))^p;
      u[i]=sin(pi*z[i]);
    }
#
#     Definition of approximations
    igrid[1]=1;  igrid[2]=2;
    igrid[n-1]=4;igrid[n]=5;
    for(i in 3:(n-2)){
      igrid[i]=3;
    }
#
#     Numerical uz, uzz
      uz=dss032(ncall,npts,igrid,n,z,u);
      uzz=dss032(ncall,npts,igrid,n,z,uz);
#
#     Exact uzz
      cat(sprintf("\n      i  igrid      z          u
        uzz        uzze       diff"));
```

```
      for(i in 1:n){
      uzze[i]=-pi^2*sin(pi*z[i]);
      diff[i]=uzz[i]-uzze[i];
#
#     Display numerical and exact derivatives, difference
      cat(sprintf("\n %5d %5d %8.3f %8.3f %8.3f %8.3f %10.5f",
        i,igrid[i],z[i],u[i],uzz[i],uzze[i],diff[i]));
      }
#
# nfcn=2 complete
  }
```

Listing A.10.2: Test program for dss032

We can note the following details about Listing A.10.2.

- After previous workspaces are cleared, and dss032, dss32a are accessed (as discussed for Listing A.10.1), the problem and grid parameters are defined numerically.

```
#
# Problem, grid parameters
  npts=5;n=21;nfcn=1;ncall=1;
```

Specifically,
  - **npts=5**: Number of points in the differentiation formula of eq. (A.10.2).
  - **n=21**: Total number of points in the spatial grid.
  - **nfcn=1**: Function to be differentiated.
  - **ncall=1**: First call to dss032 so that the weighting coefficients (e.g., in eq. (A.10.3)) are calculated for subsequent use with dss032.
- Arrays for the numerical testing are declared (preallocated).

```
#
# Arrays for test problem
  z=rep(0,n);igrid=rep(0,n);
  u=rep(0,n);uzze=rep(0,n);diff=rep(0,n);
```

- Selected parameters are displayed at the beginning of the output.

```
#
# Display selected parameters
    cat(sprintf("\n nfcn = %2d   n = %3d\n\n",nfcn,n));
```

- A uniform grid ($p = 1$) or nonuniform grid ($p \neq 1$) is specified. Then the grid is defined (z) and a polynomial is selected (u[i]) by deactivating a comment. In this case, a fourth-order polynomial is selected.

```
#
# Polynomial for varying order (selected with #)
   if(nfcn==1){
     p=1.5;
     for(i in 1:n){
       z[i]=(0.05*(i-1))^p;
#      u[i]=1;
#      u[i]=z[i];
#      u[i]=z[i]^2;
#      u[i]=z[i]^3;
       u[i]=z[i]^4;
#      u[i]=z[i]^5;
#      u[i]=z[i]^6;
     }
```

The polynomials are of order 0 (a constant) to 6.

- The approximation at each of the n grid points is defined.

```
#
#    Definition of approximations
     igrid[1]=1;  igrid[2]=2;
     igrid[n-1]=4;igrid[n]=5;
     for(i in 3:(n-2)){
        igrid[i]=3;
     }
```

Eq. (A.10.2) is used at

- igrid[1]=1: The left end of the grid with $z = z[1]$.
- igrid[2]=2: One point from the left end of the grid with $z = z[2]$.
- igrid[n-1]=4: One point from the right end of the grid with $z = z[n-1]$.
- igrid[n]=5: The right end of the grid with $z = z[n]$.
- igrid[i]=3: The intermediate points with $z = z[3],z[4],\ldots,z[n-2]$.

For the uniform grid case of eqs. (A.10.5) (p=1 in Listing A.10.2), igrid corresponds to the rows of the differentiation matrix of eq. (A.10.7).

igrid[1]	-50	96	-72	32	-6
igrid[2]	-6	-20	36	-12	2
igrid[3] to					
igrid[n-2]	2	-16	0	16	-2
igrid[n-1]	-2	12	-36	20	6
igrid[n]	6	-32	72	-96	50

- uz is computed by dss032, and uzz by dss032 called a second time.

```
#
#    Numerical uz, uzz
     uz=dss032(ncall,npts,igrid,n,x,u);
     uzz=dss032(ncall,npts,igrid,n,z,uz);
```

- The exact value of uzz is computed by analytically differentiating the selected polynomial. For the fourth-order polynomial $u_4(z) = u^4$, $d^2u_4(z)/dz^2 = 12u^2$.

```
#
#    Exact uzz (selected with #)
     cat(sprintf("\n    i   igrid     z        u
         uzz      uzze      diff"));
     for(i in 1:n){
#    uzze[i]=0;
#    uzze[i]=2;
#    uzze[i]=6*z[i];
     uzze[i]=12*z[i]^2;
#    uzze[i]=20*z[i]^3;
#    uzze[i]=30*z[i]^4;
     diff[i]=uzz[i]-uzze[i];
```

- The numerical uzz, exact uzz, and the difference are displayed as a function of z.

```
#
#    Display numerical and exact derivatives, difference
     cat(sprintf("\n %5d %5d %8.3f %8.3f %8.3f %8.3f %10.5f",
         i,igrid[i],z[i],u[i],uzz[i],uzze[i],diff[i]));
     }
#
# nfcn=1 complete
  }
```

nfcn=1 for the polynomial test functions is then concluded.
- A similar set of statements follows for nfcn=2 ($\sin(z)$).

```
#
# Sine function
  if(nfcn==2){
    p=1.5
    for(i in 1:n){
      z[i]=(0.05*(i-1))^p;
      u[i]=sin(pi*z[i]);
    }
```

```
#
#    Exact uzz
     cat(sprintf("\n      i  igrid      z        u
       uzz       uzze      diff"));
     for(i in 1:n){
     uzze[i]=-pi^2*sin(pi*z[i]);
     diff[i]=uzz[i]-uzze[i];
```

This concludes the discussion of the programming for the test functions. Representative output follows.

Abbreviated output for ifcn=1, p=1.5 (polynomial test functions, variable grid) and a zeroth-order polynomial (a constant) follows. The grid spacing at $z = 0$ is $0.011 - 0.000 = 0.011$ and at $z = 1$, $1.000 - 0.926 = 0.074$, reflecting the variable grid. As expected, the approximation of eq. (A.10.2) is exact. However, this test is worthwhile since it might reveal a programming error.

Abbreviated output for ifcn=1, p=1.5 (polynomial test functions, variable grid) and a fourth-order polynomial follows. The complete output indicates that diff=0 so that eqs. (A.10.1) and (A.10.2) are exact. This is expected since $f(z)$ in eq. (A.10.1) is a fourth-order polynomial and generally a nth-order polynomial based on $n + 1$ distinct points is unique.

Abbreviated output for ifcn=1, p=1.5 (polynomial test functions, variable grid) and a fifth-order polynomial follows. Eqs. (A.10.1) and (A.10.2) are not exact (diff is nonzero) as expected since $f(z)$ in eq. (A.10.1) is a fourth-order polynomial while the test function is a fifth-order polynomial. The same conclusion follows for polynomials of order greater than five (this can be tested by using the sixth-order polynomial in Listing A.10.2).

i	igrid	z	u	uzz	uzze	diff
1	1	0.000	1.000	0.000	0.000	0.00000
2	2	0.011	1.000	0.000	0.000	0.00000
3	3	0.032	1.000	-0.000	0.000	-0.00000
		.			.	
		.			.	
		.			.	
	Output for i = 4,...,18 removed					
		.			.	
		.			.	
		.			.	
19	3	0.854	1.000	0.000	0.000	0.00000
20	4	0.926	1.000	-0.000	0.000	-0.00000
21	5	1.000	1.000	-0.000	0.000	-0.00000

Table A.10.3: Abbreviated numerical output for ifcn=1, p=1.5, zeroth-order polynomial

i	igrid	z	u	uzz	uzze	diff
1	1	0.000	0.000	0.000	0.000	0.00000
2	2	0.011	0.000	0.002	0.002	0.00000
3	3	0.032	0.000	0.012	0.012	-0.00000
.		.			.	
.		.			.	
.		.			.	

Output for i = 4,...,18 removed

.		.			.	
.		.			.	
.		.			.	
19	3	0.854	0.531	8.748	8.748	0.00000
20	4	0.926	0.735	10.288	10.289	-0.00000
21	5	1.000	1.000	12.000	12.000	-0.00000

Table A.10.4: Abbreviated numerical output for ifcn=1, p=1.5, fourth-order polynomial

i	igrid	z	u	uzz	uzze	diff
1	1	0.000	0.000	0.000	0.000	0.00047
2	2	0.011	0.000	0.000	0.000	0.00006
3	3	0.032	0.000	0.000	0.001	-0.00014
.		.			.	
.		.			.	
.		.			.	

Output for i = 4,...,18 removed

.		.			.	
.		.			.	
.		.			.	
19	3	0.854	0.454	12.451	12.449	0.00284
20	4	0.926	0.681	15.879	15.878	0.00141
21	5	1.000	1.000	19.969	20.000	-0.03095

Table A.10.5: Abbreviated numerical output for ifcn=1, p=1.5, fifth-order polynomial

Also, the largest value of diff is at the boundary z = 1. This is generally to be expected, that is, the noncentered approximations[5] at or near the boundaries

---

[5] $df(z_1)/dz$ of eq. (A.10.3) is a function of $f(z_1), f(z_2), f(z_3), f(z_4), f(z_5)$ (the values of $f(z)$ at z_1 and to the right of z_1), so the approximation is not centered on z_1 (is noncentered). Similarly, $df(z_2)/dz$ from eq. (A.10.2) with $z = z_2$ is a function of $f(z_1), f(z_2), f(z_3), f(z_4), f(z_5)$ (one value to the left of z_2 and three values to the right of z_2), so it is not centered on z_2. The same conclusions apply when $z = z_4, z_5$ are used in eq. (A.10.2) (noncentered approximations of $df(z)/dz$ result). For the interior point $z = z_3$, the approximation of $df(z_3)/dz$ from eq. (A.10.2) is a function of $f(z_1), f(z_2), f(z_3), f(z_4), f(z_5)$ (values of $f(z)$ centered around $z = z_3$). For a grid of $n > 5$ points, the approximation of $df(z)/dz$ from eq. (A.10.2) for the interior points $z_3, \cdot, z_i, \cdot, z_{n-2}$ is centered on z_i (it is based on $f(z_{i-2}), f(z_{i-1}), f(z_i), f(z_{i+1}), f(z_{i+2})$).

i	igrid	z	u	uzz	uzze	diff
1	1	0.000	0.000	0.001	-0.000	0.00119
2	2	0.011	0.035	-0.346	-0.347	0.00014
3	3	0.032	0.099	-0.979	-0.979	-0.00036

Output for i = 4,...,18 removed

i	igrid	z	u	uzz	uzze	diff
19	3	0.854	0.443	-4.381	-4.375	-0.00601
20	4	0.926	0.231	-2.279	-2.275	-0.00378
21	5	1.000	0.000	0.075	-0.000	0.07456

Table A.10.6: Abbreviated numerical output for ifcn=2, p=1.5

(igrid[1]=1, igrid[2]=2, igrid[n-1]=4, igrid[n]=5) will have a larger error than the centered approximations at the interior points (igrid[i]=3,i=3,...,n-2). The relatively small error at z = 0 results from the smaller grid spacing at this boundary (and also, the fifth-order polynomial test function varies more rapidly at z=1 than at z=0).

Finally, abbreviated output for ifcn=2, p=1.5 (sine test function, variable grid) is in Table A.10.6. Eqs. (A.10.1) and (A.10.2) are not exact (diff is nonzero) as expected since the sine function can be considered as a polynomial of infinite order (consider its Taylor series expansion). The largest value of diff is at the boundary z = 1. The relatively small error at z = 0 results from the smaller grid spacing at this boundary.

In conclusion, approximations such as eqs. (A.10.1) and (A.10.2) based on the Lagrange interpolation polynomial are available for derivatives of any order, and any number of grid points, with uniform or variable spacing [2]. Thus, approximations of spatial derivatives in PDEs can be constructed under very general conditions to meet the requirements of a particular PDE application.

References

[1] Bickley, W.G. (1941), Formulae for numerical differentiation, *Math. Gaz.*, **25**, 263, p 22

[2] Fornberg, B. (1991), *Recent Developments in Numerical Methods and Software for ODEs/DAEs/PDEs*, Byrne, G.D. and W.E. Schiesser, eds., World Scientific, Singapore, pp 97–123

[3] Schiesser, W.E. (2012), *Partial Differential Analysis in Biomedical Engineering*, Chapter 2, Cambridge University Press, Cambridge, UK

[4] Schiesser, W.E. (2014), *Differential Equation Analysis in Biomedical Science and Engineering; Partial Differential Equation Analysis in R*, John Wiley, Hoboken, NJ

[5] Vijayendran, R.A., F.S. Ligler, and D.E. Leckband (1999), A computational reaction-diffusion model for the analysis of transport-limited kinetics, *Anal. Chem.*, **71**, 5405–5412

APPENDIX A

DERIVATION OF CONVECTION-DIFFUSION-REACTION PARTIAL DIFFERENTIAL EQUATIONS

A mass balance in cylindrical coordinates (r, θ, z) on an incremental volume $(r\Delta\theta)(\Delta r)(\Delta z)$[1] ([1], p 840) gives

$$(r\Delta\theta)(\Delta r)(\Delta z)\frac{\partial c}{\partial t} = \tag{A.1.1}$$

$$(r\Delta\theta)(\Delta z)v_r c|_r - (r\Delta\theta)(\Delta z)v_r c|_{r+\Delta r} \tag{A.1.2}$$

$$+(\Delta r)(\Delta z)v_\theta c|_\theta - (\Delta r)(\Delta z)v_\theta c|_{\theta+\Delta\theta} \tag{A.1.3}$$

$$+(r\Delta\theta)(\Delta r)v_z c|_z - (r\Delta\theta)(\Delta r)v_z c|_{z+\Delta z} \tag{A.1.4}$$

$$-(r\Delta\theta)(\Delta z)D_{rr}\frac{\partial c}{\partial r}\Big|_r - \left(-(r\Delta\theta)(\Delta z)D_{rr}\frac{\partial c}{\partial r}\Big|_{r+\Delta r}\right) \tag{A.1.5}$$

$$-(r\Delta\theta)(\Delta z)D_{r\theta}\frac{\partial c}{r\partial\theta}\Big|_r - \left(-(r\Delta\theta)(\Delta z)D_{r\theta}\frac{\partial c}{r\partial\theta}\Big|_{r+\Delta r}\right) \tag{A.1.6}$$

$$-(r\Delta\theta)(\Delta z)D_{rz}\frac{\partial c}{\partial z}\Big|_r - \left(-(r\Delta\theta)(\Delta z)D_{rz}\frac{\partial c}{\partial z}\Big|_{r+\Delta r}\right) \tag{A.1.7}$$

$$-(\Delta r)(\Delta z)D_{\theta r}\frac{\partial c}{\partial r}\Big|_\theta - \left(-(\Delta r)(\Delta z)D_{\theta r}\frac{\partial c}{\partial r}\Big|_{\theta+\Delta\theta}\right) \tag{A.1.8}$$

---

[1]c is used for the dependent variable since the mass balance is typically for the concentration of a chemical component. Individual terms in the balance are numbered to facilitate referring to the terms.

Method of Lines PDE Analysis in Biomedical Science and Engineering, First Edition. William E. Schiesser.
© 2016 John Wiley & Sons, Inc. Published 2016 by John Wiley & Sons, Inc.
Companion website: www.wiley.com/go/Schiesser/PDE_Analysis

$$-(\Delta r)(\Delta z)D_{\theta\theta}\frac{\partial c}{r\partial\theta}|_\theta - \left(-(\Delta r)(\Delta z)D_{\theta\theta}\frac{\partial c}{r\partial\theta}|_{\theta+\Delta\theta}\right) \tag{A.1.9}$$

$$-(\Delta r)(\Delta z)D_{\theta z}\frac{\partial c}{\partial z}|_\theta - \left(-(\Delta r)(\Delta z)D_{\theta z}\frac{\partial c}{\partial z}|_{\theta+\Delta\theta}\right) \tag{A.1.10}$$

$$-(r\Delta\theta)(\Delta r)D_{zr}\frac{\partial c}{\partial r}|_z - \left(-(r\Delta\theta)(\Delta r)D_{zr}\frac{\partial c}{\partial r}|_{z+\Delta z}\right) \tag{A.1.11}$$

$$-(r\Delta\theta)(\Delta r)D_{z\theta}\frac{\partial c}{r\partial\theta}|_z - \left(-(r\Delta\theta)(\Delta r)D_{z\theta}\frac{\partial c}{r\partial\theta}|_{z+\Delta z}\right) \tag{A.1.12}$$

$$-(r\Delta\theta)(\Delta r)D_{zz}\frac{\partial c}{\partial z}|_z - \left(-(r\Delta\theta)(\Delta r)D_{zz}\frac{\partial c}{\partial z}|_{z+\Delta z}\right) \tag{A.1.13}$$

$$+(r\Delta\theta)(\Delta r)(\Delta z)Q_r(c) \tag{A.1.14}$$

where

- c: PDE-dependent variable, i.e., $c(r,\theta,z,t)$.
- r,θ,z: cylindrical spatial coordinates.
- t: time.
- v_r, v_θ, v_z: components of the velocity vector $\mathbf{v} = \mathbf{i}_r v_r + \mathbf{j}_\theta v_\theta + \mathbf{k}_z v_z$. $\mathbf{i}_r, \mathbf{j}_\theta, \mathbf{k}_z$ are the components of the orthonormal (unit) vector in cylindrical coordinates.
- $D_{rr}, \ldots, D_{zz}$: components of the diffusivity tensor

$$\mathbf{D} = \begin{bmatrix} D_{rr} & D_{r\theta} & D_{rz} \\ D_{\theta r} & D_{\theta\theta} & D_{\theta z} \\ D_{zr} & D_{z\theta} & D_{zz} \end{bmatrix}$$

- $Q_r(c)$: volumetric rate of reaction. $Q_r(c) < 0$ corresponds to consumption of chemical reactant; $Q_r(c) > 0$ corresponds to production of chemical product.

Rearrangement (division by $(r\Delta\theta)(\Delta r)(\Delta z)$) of eq. (A.1) gives

$$\frac{\partial c}{\partial t} = \tag{A.2.1}$$

$$-\frac{r\Delta\theta\Delta z v_r c|_{r+\Delta r} - r\Delta\theta\Delta z v_r c|_r}{r\Delta\theta\Delta r\Delta z} \tag{A.2.2}$$

$$-\frac{\Delta r\Delta z v_\theta c|_{\theta+\Delta\theta} - \Delta r\Delta z v_\theta c|_\theta}{r\Delta\theta\Delta r\Delta z} \tag{A.2.3}$$

$$-\frac{r\Delta\theta\Delta r v_z c|_{z+\Delta z} - (r\Delta\theta)(\Delta r)v_z c|_z}{r\Delta\theta\Delta r\Delta z} \tag{A.2.4}$$

$$\frac{r\Delta\theta\Delta z D_{rr}\frac{\partial c}{\partial r}|_{r+\Delta r} - r\Delta\theta\Delta z D_{rr}\frac{\partial c}{\partial r}|_r}{r\Delta\theta\Delta r\Delta z} \tag{A.2.5}$$

$$+\frac{r\Delta\theta\Delta z D_{r\theta}\frac{\partial c}{r\partial\theta}|_{r+\Delta r} - r\Delta\theta\Delta z D_{r\theta}\frac{\partial c}{r\partial\theta}|_r}{r\Delta\theta\Delta r\Delta z} \tag{A.2.6}$$

$$+\frac{r\Delta\theta\Delta z D_{rz}\frac{\partial c}{\partial z}|_{r+\Delta r} - r\Delta\theta\Delta z D_{rz}\frac{\partial c}{\partial z}|_r}{r\Delta\theta\Delta r\Delta z} \tag{A.2.7}$$

$$+\frac{\Delta r\Delta z D_{\theta r}\frac{\partial c}{\partial r}|_{\theta+\Delta\theta} - \Delta r\Delta z D_{\theta r}\frac{\partial c}{\partial r}|_\theta}{r\Delta\theta\Delta r\Delta z} \tag{A.2.8}$$

$$+\frac{\Delta r\Delta z D_{\theta\theta}\frac{\partial c}{r\partial\theta}|_{\theta+\Delta\theta} - \Delta r\Delta z D_{\theta\theta}\frac{\partial c}{r\partial\theta}|_\theta}{r\Delta\theta\Delta r\Delta z} \tag{A.2.9}$$

$$+\frac{\Delta r\Delta z D_{\theta z}\frac{\partial c}{\partial z}|_{\theta+\Delta\theta} - \Delta r\Delta z D_{\theta z}\frac{\partial c}{\partial z}|_\theta}{r\Delta\theta\Delta r\Delta z} \tag{A.2.10}$$

$$+\frac{r\Delta\theta\Delta r D_{zr}\frac{\partial c}{\partial r}|_{z+\Delta z} - r\Delta\theta\Delta r D_{zr}\frac{\partial c}{\partial r}|_z}{r\Delta\theta\Delta r\Delta z} \tag{A.2.11}$$

$$+\frac{r\Delta\theta\Delta r D_{z\theta}\frac{\partial c}{r\partial\theta}|_{z+\Delta z} - r\Delta\theta\Delta r D_{z\theta}\frac{\partial c}{r\partial\theta}|_z}{r\Delta\theta\Delta r\Delta z} \tag{A.2.12}$$

$$+\frac{r\Delta\theta\Delta r D_{zz}\frac{\partial c}{\partial z}|_{z+\Delta z} - r\Delta\theta\Delta r D_{zz}\frac{\partial c}{\partial z}|_z}{r\Delta\theta\Delta r\Delta z} \tag{A.2.13}$$

$$+Q_r(c) \tag{A.2.14}$$

In the limit $\Delta r, \Delta\theta, \Delta z \to 0$, eq. (A.2) becomes

$$\frac{\partial c}{\partial t} = \tag{A.3.1}$$

$$-\frac{1}{r}\frac{\partial(rv_r c)}{\partial r} - \frac{1}{r}\frac{\partial(v_\theta c)}{\partial\theta} - \frac{\partial(v_z c)}{\partial z} \tag{A.3.2,3,4}$$

$$+\frac{\partial\left(rD_{rr}\frac{\partial c}{\partial r}\right)}{r\partial r} + \frac{\partial\left(rD_{r\theta}\frac{\partial c}{r\partial\theta}\right)}{r\partial r} + \frac{\partial\left(rD_{rz}\frac{\partial c}{\partial z}\right)}{r\partial r} \tag{A.3.5,6,7}$$

$$+\frac{\partial\left(D_{\theta r}\frac{\partial c}{\partial r}\right)}{r\partial\theta} + \frac{\partial\left(D_{\theta\theta}\frac{\partial c}{r\partial\theta}\right)}{r\partial\theta} + \frac{\partial\left(D_{\theta z}\frac{\partial c}{\partial z}\right)}{r\partial\theta} \tag{A.3.8,9,10}$$

$$+\frac{\partial\left(D_{zr}\frac{\partial c}{\partial r}\right)}{\partial z} + \frac{\partial\left(D_{z\theta}\frac{\partial c}{r\partial\theta}\right)}{\partial z} + \frac{\partial\left(D_{zz}\frac{\partial c}{\partial z}\right)}{\partial z} \tag{A.3.11,12,13}$$

$$+Q_R(c) \tag{A.3.14}$$

For constant velocity and diffusivity, eq. (A.3) becomes

$$\frac{\partial c}{\partial t} = \tag{A.4.1}$$

$$-\frac{v_r}{r}\frac{\partial(rc)}{\partial r} - \frac{v_\theta}{r}\frac{\partial c}{\partial \theta} - v_z\frac{\partial c}{\partial z} \tag{A.4.2,3,4}$$

$$+D_{rr}\left(\frac{\partial^2 c}{\partial r^2} + \frac{1}{r}\frac{\partial c}{\partial r}\right) + \frac{D_{r\theta}}{r}\frac{\partial^2 c}{\partial r\partial\theta} + D_{rz}\left(\frac{\partial^2 c}{\partial r\partial z} + \frac{1}{r}\frac{\partial c}{\partial z}\right) \tag{A.4.5,6,7}$$

$$+\frac{D_{\theta r}}{r}\frac{\partial^2 c}{\partial\theta\partial r} + \frac{D_{\theta\theta}}{r^2}\frac{\partial^2 c}{\partial\theta^2} + \frac{D_{\theta z}}{r}\frac{\partial^2 c}{\partial\theta\partial z} \tag{A.4.8,9,10}$$

$$+D_{zr}\frac{\partial^2 c}{\partial z\partial r} + \frac{D_{z\theta}}{r}\frac{\partial^2 c}{\partial z\partial\theta} + D_{zz}\frac{\partial^2 c}{\partial z^2} \tag{A.4.11,12,13}$$

$$+Q_r(c) \tag{A.4.14}$$

With zero off-diagonal elements of the diffusivity tensor, eq. (A.4) becomes

$$\frac{\partial c}{\partial t} = -\frac{v_r}{r}\frac{\partial(rc)}{\partial r} - \frac{v_\theta}{r}\frac{\partial c}{\partial \theta} - v_z\frac{\partial c}{\partial z}$$
$$+ D_{rr}\left(\frac{\partial^2 c}{\partial r^2} + \frac{1}{r}\frac{\partial c}{\partial r}\right) + \frac{D_{\theta\theta}}{r^2}\frac{\partial^2 c}{\partial\theta^2} + D_{zz}\frac{\partial^2 c}{\partial z^2} + Q_r(c) \tag{A.5}$$

Eq. (A.5) is the starting point for the discussion of first order hyperbolic (convection) PDEs in Chapter 1, for the special case of the linear advection equation ($v_r = v_\theta = D_{rr} = D_{\theta\theta} = D_{zz} = Q_r(c) = 0$)

$$\frac{\partial c}{\partial t} = -v_z\frac{\partial c}{\partial z} \tag{A.6}$$

Reference

[1] Bird, R.B., W.E. Stewart, and E.N. Lightfoot (2002), *Transport Phenomena*, 2nd edn, John Wiley & Sons, Inc., New York.

APPENDIX B

FUNCTIONS DSS012, DSS004, DSS020, VANL

Listings of functions dss012, dss004, dss020, vanl follow.

(B1) Function dss012

```
  dss012=function(xl,xu,n,u,v) {
#
# Function dss012 computes the first order finite difference
# approximation of a first derivative
#
# Declare arrays
  ux=rep(0,n);
#
# Grid spacing
  dx=(xu-xl)/(n-1);
#
# Finite difference approximation for positive v
  if(v > 0){
    ux[1]=(u[2]-u[1])/dx;
    for(i in 2:n){
      ux[i]=(u[i]-u[i-1])/dx;}
  }
#
# Finite difference approximation for negative v
```

```
  if(v < 0){
    for(i in 1:(n-1)){
      ux[i]=(u[i+1]-u[i])/dx;}
    ux[n]=(u[n]-u[n-1])/dx;
    }
#
# All points concluded (x=xl,...,x=xu)
  return(c(ux));
}
```

(B2) Function dss004

```
  dss004=function(xl,xu,n,u) {
#
# An extensive set of documentation comments detailing the
# derivation of the following fourth order finite differences
# (FDs) is not given here to conserve space.  The derivation is
# detailed in Schiesser, W. E., The Numerical Method of Lines
# Integration of Partial Differential Equations, Academic Press,
# San Diego, 1991.
#
# Preallocate arrays
  ux=rep(0,n);
#
# Grid spacing
  dx=(xu-xl)/(n-1);
#
# 1/(12*dx) for subsequent use
  r12dx=1/(12*dx);
#
# ux vector
#
# Boundaries (x=xl,x=xu)
  ux[1]=r12dx*(-25*u[1]+48*u[ 2]-36*u[ 3]+16*u[ 4]-3*u[ 5]);
  ux[n]=r12dx*( 25*u[n]-48*u[n-1]+36*u[n-2]-16*u[n-3]+3*u[n-4]);
#
# dx in from boundaries (x=xl+dx,x=xu-dx)
  ux[ 2]=r12dx*(-3*u[1]-10*u[ 2]+18*u[ 3]-6*u[ 4]+u[ 5]);
  ux[n-1]=r12dx*( 3*u[n]+10*u[n-1]-18*u[n-2]+6*u[n-3]-u[n-4]);
#
# Interior points (x=xl+2*dx,...,x=xu-2*dx)
  for(i in 3:(n-2))ux[i]=r12dx*(-u[i+2]+8*u[i+1]-8*u[i-1]+u[i-2]);
#
# All points concluded (x=xl,...,x=xu)
  return(c(ux));
}
```

(B3) Function dss020

```
   dss020=function(xl,xu,n,u,v) {
#
# An extensive set of documentation comments detailing the
# derivation of the following five-point upwind finite differences
# (FDs) is not given here to conserve space.  The derivation is
# detailed in Schiesser, W. E., The Numerical Method of Lines
# Integration of Partial Differential Equations, Academic Press,
# San Diego, 1991.
#
# Declare arrays
   ux=rep(0,n);
#
# Grid spacing
#
# 1/(12*dx) for subsequent use
   dx=(xu-xl)/(n-1);
   r12dx=1/(12*dx);
#
# (1)  Finite difference approximation for positive v
#
   if(v>0){
   ux[1]=r12dx*(-25*u[1]+48*u[2]-36*u[3]+16*u[4] -3*u[5]);
   ux[2]=r12dx*( -3*u[1]-10*u[2]+18*u[3] -6*u[4]    +u[5]);
   ux[3]=r12dx*(    u[1] -8*u[2]+             +8*u[4]  -u[5]);
   for(i in 4:(n-1)){
   ux[i]=r12dx*(   -u[i-3] +6*u[i-2]-18*u[i-1]+10*u[i] +3*u[i+1]);}
   ux[n]=r12dx*(   3*u[n-4]-16*u[n-3]+36*u[n-2]-48*u[n-1]+25*u[n]);
   }
#
# (2)  Finite difference approximation for negative v
#
   if(v<0){
   ux[1]=r12dx*( -25*u[1]+48*u[2]-36*u[3]+16*u[4] -3*u[5]);
   for(i in 2:(n-3)){
   ux[i]=r12dx*(  -3*u[i-1]-10*u[i]+18*u[i+1] -6*u[i+2]   +u[i+3]);}
   ux[n-2]=r12dx*(   u[n-4] -8*u[n-3]           +8*u[n-1]   -u[n]);
   ux[n-1]=r12dx*(  -u[n-4] +6*u[n-3]-18*u[n-2]+10*u[n-1] +3*u[n]);
   ux[n]=r12dx*(   3*u[n-4]-16*u[n-3]+36*u[n-2]-48*u[n-1]+25*u[n]);
   }
#
# All points concluded (x=xl,...,x=xu)
   return(c(ux));
}
```

(B4) Function vanl

The van Leer flux limiter programmed in the following routine is one of a series of limiters discussed in [1], pp. 37–43. These limiters have a common form, so they can be programmed by a straightforward modification of phi in the following code. Examples include the Superbee and Smart limiters in routines super and smart discussed in Chapter 1.

```
  vanl=function(xl,xu,n,u,v) {
#
# Function vanl computes the first order finite difference
# (flux limiter) approximation of a first derivative
#
# Declare arrays
  ux=rep(0,n)
  phi=rep(0,n)
  r=rep(0,n)
#
# Grid spacing
  dx=(xu-xl)/(n-1)
#
# Tolerance for limiter switching
  delta=1.0e-05
#
# Positive v
  if(v >= 0){
    for(i in 3:(n-1)){
      if(abs(u[i]-u[i-1])<delta)
        phi[i]=0
      else{
        r[i]=(u[i+1]-u[i])/(u[i]-u[i-1])
        if(r[i]<0)
          phi[i]=0
        else
        phi[i]=max(0,min(2*r[i],min(0.5*(1.0+r[i]),2)))}
      if(abs(u[i-1]-u[i-2])<delta)
        phi[i-1]=0
      else{
        r[i-1]=(u[i]-u[i-1])/(u[i-1]-u[i-2])
        if(r[i-1]<0)
          phi[i-1]=0
        else
        phi[i-1]=max(0,min(2*r[i-1],min(0.5*(1.0+r[i-1]),2)))}
      flux2=u[i  ]+(u[i  ]-u[i-1])*phi[i  ]/2
      flux1=u[i-1]+(u[i-1]-u[i-2])*phi[i-1]/2
      ux[i]=(flux2-flux1)/dx
```

```
    }
      ux[1]=(-u[1]+u[2])/dx
      ux[2]=(-u[1]+u[2])/dx
      ux[n]=(u[n]-u[n-1])/dx
  }
#
# Negative v
  if(v < 0){
    for(i in 2:(n-2)){
      if(abs(u[i]-u[i+1])<delta)
         phi[i]=0
      else{
         r[i]=(u[i-1]-u[i])/(u[i]-u[i+1])
         if(r[i]<0)
           phi[i]=0
         else
         phi[i]=max(0,min(2*r[i],min(0.5*(1.0+r[i]),2)))}
      if(abs(u[i+1]-u[i+2])<delta)
         phi[i+1]=0
      else{
         r[i+1]=(u[i]-u[i+1])/(u[i+1]-u[i+2])
         if(r[i+1]<0)
           phi[i+1]=0
         else
           phi[i+1]=max(0,min(2*r[i+1],min(0.5*(1.0+r[i+1]),2)))}
      flux2=u[i  ]+(u[i  ]-u[i+1])*phi[i  ]/2
      flux1=u[i+1]+(u[i+1]-u[i+2])*phi[i+1]/2
      ux[i]=-(flux2-flux1)/dx
    }
      ux[1]=(-u[1]+u[2])/dx
      ux[n-1]=(-u[n-1]+u[n])/dx
      ux[n]  =(-u[n-1]+u[n])/dx
  }
#
# All points concluded (x=xl,...,x=xu)
  return(c(ux))
}
```

Reference

[1] Griffiths, G.W., and W.E. Schiesser (2012), *Traveling Wave Analysis of Partial Differential Equations*, Elsevier/Academic Press, Boston, MA.

INDEX

Method of Lines PDE Analysis in Biomedical Science and Engineering, First Edition. William E. Schiesser.
© 2016 John Wiley & Sons, Inc. Published 2016 by John Wiley & Sons, Inc.
Companion website: www.wiley.com/go/Schiesser/PDE_Analysis